Tribology in Chemical-Mechanical Planarization

Tribology in Chemical-Mechanical Planarization

Hong Liang
David Craven

CRC Press
Taylor & Francis Group
Boca Raton London New York

CRC Press is an imprint of the
Taylor & Francis Group, an **informa** business

A TAYLOR & FRANCIS BOOK

CRC Press
Taylor & Francis Group
6000 Broken Sound Parkway NW, Suite 300
Boca Raton, FL 33487-2742

First issued in paperback 2019

© 2005 by Taylor & Francis Group, LLC
CRC Press is an imprint of Taylor & Francis Group, an Informa business

No claim to original U.S. Government works

ISBN-13: 978-0-8247-2567-9 (hbk)
ISBN-13: 978-0-367-39325-0 (pbk)

Library of Congress Card Number 2004058221

Library of Congress Cataloging-in-Publication Data

Liang, Hong, 1961-
  Tribology in chemical-mechanical planarization / Hong Liang, David R. Craven.
    p. cm.
  Includes bibliographical references and index.
  ISBN 0-8247-2567-0
  1. Tribology. 2. Grinding and polishing. I. Craven, David R., 1945- II. Title.

TJ1075.L49 2004
621.9'2--dc22   2004058221

Visit the Taylor & Francis Web site at
http://www.taylorandfrancis.com

and the CRC Press Web site at
http://www.crcpress.com

Preface

This book has two objectives. First, we seek to present the fundamentals of tribology to those scientists and engineers engaged in chemical-mechanical planarization (CMP) who want to extend their knowledge in this important science and technology. Likewise, the book attempts to provide an overview of the CMP process for tribology professionals who are interested in expanding their expertise in this fast-growing technology. The field of tribology has existed for many years and has played an important role in manufacturing technology development. CMP technology is still young, and much of the industrial positioning is based on empirical explorations. In time, we hope, the tribological principles that underlie CMP will play more important roles in the development of the technology's future.

This book is divided into eight chapters. Chapter 1 is an introduction to tribology and CMP. A brief history of tribology is provided here to highlight the important roles this field has played in manufacturing technology development. The historical development of integrated circuits and CMP are introduced in the second half of this chapter.

Chapters 2 through 5 are focused on tribological principles and their applications in CMP. Chapter 2 discusses the surfaces of contacts, involving basic concepts such as surface roughness, conformity, real contact area, and wear–contact mechanism. At the end of this chapter, common defects in CMP are discussed. Chapter 3 focuses on the basics of friction as well as flash temperature and friction in CMP. Chapter 4 discusses lubrication fundamentals and their roles in CMP.

Examples and guidelines for CMP are given in later chapters: Chapter 5 focuses on wear. The basics of wear, polishing particles, and pad wear are discussed in this chapter. Finally, Chapters 6, 7, and 8 deal with CMP practices. Because forces are important factors linking polishing performance and tribology, Chapter 6 discusses these particular mechanical aspects of CMP. Pad materials, i.e., polymers, are discussed for their viscoelastic properties as well as elastic modulus and cell buckling. Chapter 7 deals with pads and Chapter 8 with cleaning technology.

Because this book will serve at least two communities — tribology and CMP — it will be beneficial for chemists, materials scientists, electrical and mechanical engineers, physicists, tribologists, and many other kinds of applied scientists and engineers. Graduate students will learn from this book how to use classroom knowledge to understand cutting-edge technology.

We wish to express our gratitude to the staff at Marcel Dekker Incorporated and at CRC Press for their invaluable assistance. Liang's special thanks to George Totten and Jorn Larsen-Bassefor their continuous encouragement. Her thanks also go to her colleagues, postdoctorals, and students who stand by her. Craven would like to dedicate this book to his parents who have been a constant source of inspiration and support.

Hong Liang
David Craven

About the Authors

Hong Liang is associate professor in the Department of Mechanical Engineering at Texas A&M University. She has been a visiting professor at the Department of Materials Science and Engineering at the Massachusetts Institute of Technology, the Energy Technology Division at Argonne National Laboratory, and the Tribology and Dynamics Laboratory at the École Centrale de Lyon in France. The author or coauthor of more than 60 professional publications, she is the recipient of the AFM International Lectureship Award (1999) and the NSF Career Award (2002). She is a member of the American Society of Materials International, the American Association of Mechanical Engineers, the Materials Research Society, and the Society of Tribologists and Lubrication Engineers. She received a B.S. degree in materials science and engineering from Beijing University of Iron and Steel Technology in China, and M.S. and Ph.D. degrees in materials science and engineering from the Stevens Institute of Technology at Hoboken, New Jersey. She completed postdoctoral research at the National Institute of Standards and Technology in Gaithersburg, Maryland.

David Craven is a senior member of the staff in the Product Knowledge Business Unit at Lam Research Corporation, where he has been involved with the semiconductor technologies of CMP and Etch. He has been on the faculty at Foothill College and at the University of Phoenix, and has taught at Northwestern University. He is currently a member of the IEEE Society. He has been working in semiconductor technology for more than 35 years and is the author of numerous papers in the field of semiconductor processing and microelectromechanical systems. He received a B.S. degree in physics from Whitman College, an M.S. degree in physicsfrom Dartmouth College, and a degree in electrical engineering from Santa Clara University.

Contents

chapter one

Introduction

In the microelectronics community, the acronym CMP has been interpreted as either chemical-mechanical polishing or chemical-mechanical planarization. Polishing means the removal of materials, whereas planarization means flattening. It is the authors' preference to interpret CMP as chemical-mechanical planarization. Because technology and integrated circuit (IC) shrinking advance rapidly, planarization has come to be at the heart of the CMP function.

Tribological investigation of CMP has started to attract the attention of industrial engineers. It includes studies of friction, wear, lubrication, contact mechanics, and tribochemistry. The wear or abrasion of a surface material requires the application of a force. Material removal is, however, a strong function of the micro- or nanospecifics of the interacting forces, so that the nature of the force transmission, from its macroapplication to its nanointeraction with the surface, is critical. In the typical scenario, CMP tools apply mechanical pressure to a wafer carrier to compress the silicon wafer against a pliable polymer pad. The behavior of this polymer under process conditions is nontrivial and often not overtly intuitive.

Once the force of interaction is engaged at the wafer surface, the elements of friction and lubrication become dominant components in the removal of surface material. Thus the role that abrasion and lubrication play in the processes of wear and planarization is central to the results of CMP technology. But at this nanoengagement level, the interactions of surface interface chemistry become enormously influential. Not only are the interface layer characteristics (thickness, composition, roughness, etc.) critical to the behavior of material removal results, but also the localized reaction rate, byproduct buildup, temperature distribution, and numerous other effects can radically shift the process into different reaction regimes that obscure a clear understanding of the precise removal mechanisms. The objective of this book is to present the fundamentals of tribology to those engineers engaged in CMP who want to extend their knowledge in this important science and technology. Likewise, the text attempts to provide an overview of the CMP process for tribology professionals who are interested in expanding their expertise in the fast-growing technology of CMP. A number of intriguing ideas are raised in various sections of this book. The validity of some of these ideas is not yet established, so we will have to wait for further research as we continue to build the scientific foundation for the science of CMP. We must apologize, in a sense, for raising more questions than we solve; but part of the purpose of this book is to provide an orientation for future work.

1.1   A brief history of tribology

The term *tribology* was first used in 1966,[1] and it is a compound of two Greek words, meaning *the science of rubbing*. It is usually understood as the science and technology of interacting surfaces in relative motion and the practices related to it. The popular associations include the notions of friction and wear, as well as lubrication. A more specific definition can be stated as: "Tribology is the art of applying operational analysis to problems of great economic significance, namely — reliability, maintenance, and wear of technical equipment."[2]

One of the earliest uses of friction can be seen in some of the Neandertal implements (Figure 1.1), where grinding had developed into an art with both social and survival implications.

We can easily imagine that ever since our ancestors began to drag loads over the ground, they developed methods to reduce frictional forces. This started as early as 3500 B.C. using wheels in translationary motion. There is evidence from ca. 1880 B.C. that Egyptian craftsmen, in transporting large stone blocks and carvings, used lubrication to reduce friction in both sliding and rolling contacts[4,5] (Figure 1.2). Dating back to the middle of the second millennium B.C., Chinese pictographs have been found showing wheeled chariots. Potters' wheels using fired porcelain cups as bearings appeared in China about 1500 B.C. By 400 B.C. Chinese bearing technology had developed sophisticated lubricated bronze bearings for use on war chariots.[4,5] During the Greco–Roman period, military engineers rose to prominence by devising both war machinery and methods of fortification using tribological principles.

The scientific approach to tribology started with Leonardo da Vinci (1452–1519), who introduced the concept of the coefficient of friction as the ratio of the friction force to a normal load (Figure 1.3). He was celebrated in his days for his genius in military construction as well as for his painting and sculpture.

Figure 1.1 Use of abrasion in Neandertal art objects.[3]

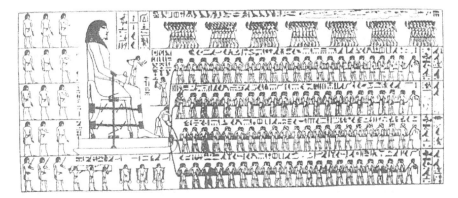

Figure 1.2 Use of lubrication by the Egyptians.[6]

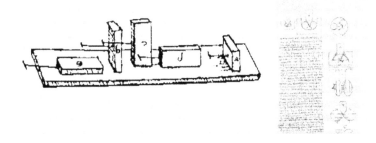

Figure 1.3 Leonardo da Vinci's work on friction.[7]

Further development in tribology occurred in 1699, a French physicist, Guillaume Amontons, rediscovered the rules of friction after he studied dry sliding between two flat surfaces.[8] He found that the friction force does not depend on the apparent area of contact. This was verified by another French physicist, Charles-Augustin Coulomb, who distinguished between static friction and kinetic friction.[9] By the 1850s, tribology had moved forward with the establishment of the petroleum industry in Scotland, Canada, and the United States.[10,1] Since the beginning of the twentieth century, enormous industrial growth led to the demand for better tribological performance, expanding the field of tribology tremendously. Today, tribological understanding has extended to the atomic level with the development of nanoscale science and technology.[11-14]

Tribology dominates our daily activities. Friction enables us to stand, walk, and run properly. Friction also enables us to drive a car, a boat, or a snowmobile or to ski or skate. Brushing, shaving, and chewing involve extensive friction, wear, and lubrication. Writing and printing are tribological processes accomplished by a controlled transfer of ink to paper. Some of these processes are productive and some are unproductive friction and wear.

1.2 CMP technology development

1.2.1 Early history

Because of the forward-thinking orientation in the IBM advanced product programs, the devices that IBM required in order to maintain its technological superiority could be seen to have future limitations as the number of metalization levels increased. Various exploratory technologies were examined in an effort to develop a manufacturable solution. It was from such work that the technology of CMP was born.[15]

The original development work on CMP was done around 1983 at East Fishkill, New York, at what was known as the Base Technology Lab. The term *Damascene*, commonly used in copper processing, is ascribed to Charles (Bud) Standley. He was aware of special jewelry techniques, which had their origins in Toledo, Spain, after the Arab invasion brought various metalworking techniques from Damascus. The process involved inlaying precious metals in grooves on a base material (often iron) and polishing the surface smooth to expose the inlay pattern. This is the essence of copper metalization today, in which a trench is etched to receive the metal, and then the transferred metal overburden is polished back with CMP techniques. CMP's early history shows the secrecy that cloaked its development and implementation. There was an early realization that advanced circuit designs were going to demand a solution to the interconnect layer limitations that would decidedly affect yield and performance. The uniqueness of the CMP solution would allow IBM to develop a substantial lead in the production of advanced integrated circuits and hence in the marketplace. The use and customization of the consumables and CMP tools were carried on without joint development supplier activities, to protect against proliferation of the technology.

Three major circumstances resulted in the exposure of CMP technology to industry. The first was the concern for quality and viability of the key components for its PC offerings in the early 1980s. These centered on Intel and Micron for the microprocessor and memory components, and prompted a sharing of interconnect technology (including CMP). Thus new centers of CMP development were created. The second was IBM's participation in SEMATECH.* By the late 1980s, IBM had agreed to disclose its multilevel interconnect technology internally to SEMATECH. This created another center for CMP development outside of IBM. In both cases, once work was under way in these centers of development, the typical employee attrition mechanisms led to the dissemination of CMP technology. The third major circumstance was the inadvertent dissemination pipeline created by IBM itself through its own employee downsizing in the early 1990s. As M.A. Fury has written, "The migration to smaller devices elevated CMP from a curious option to an immature technology on the critical path."[15]

---

* SEMATECH is an international organization involved in setting industry direction and accelerating technology solutions to ensure a strong and vibrant semiconductor industry.

1.2.2 CMP adoption

Not all integrated circuits have the line width or the layer complexity that requires CMP. Much ink is spilled on leading-edge semiconductor technology, but at the turn of the twenty-first century, only 20% of the wafers in the manufacturing process was committed to circuits using 0.25 μm technology or smaller. Quarter-micron technology is considered the point at which rigorous circuit planarity during manufacturing is required (for adequate lithographic resolution). Figure 1.4 shows the number of silicon wafers starting the manufacturing process per week.[16] It shows both the increasing percentage of wafers committed to 0.25 μm technology or smaller (a) and the number of planarization steps manufacturing will need (b). Both bar charts indicate the rapidly increasing need for CMP technology.

1.2.3 Evolution of IC and device technology

The CMOS transistor is *the* fundamental element of today's electronic devices. Thus a brief review of the design and fabrication of a representative CMOS   transistor will help us to understand the impact of CMP on the semiconductor manufacturing process. A semiconductor transistor is an active switch used to control the buildup of voltage and the flow of current from one part of a circuit to another. The transistor uses the presence of a voltage placed close to the surface of a semiconducting material to stimulate an effect in that surface material which will allow current to flow as long as the electric field from that voltage is present. This is why the transistor was originally called a field-effect transistor (FET). One of the requirements for good performance is that the silicon in the area of the field effect must be a uniform single crystal, and its surface must be very clean and smooth. A thin layer of an insulating material is grown over this area, and a conducting material (called the gate) is deposited to hold the charge voltage that can switch the transistor on or off. The transistor switching speed of the circuit will depend on how quickly the surface under the gate can be converted to allow a current to flow, and how long it takes the current to get from one side of the transistor to the other (Figure 1.5).

   If the logical computation of the circuit is to function properly, groups of transistors must be electrically isolated from each other. The early method of isolating transistors was to protect the region where the transistor was to be built and to oxidize the rest of the silicon surface, turning those local regions into silicon dioxide. This method came to be known as LOCOS for its method of locally oxidizing the silicon surface. One disadvantage of this method is that it uses up a lot of surface area during the isolation process because of the lateral extension of the oxidation process into the transistor area (forming a structure that looks somewhat like a bird's beak). A more recent method of isolating transistors is to etch a shallow trench and then deposit oxide into it for isolation. This method is known as the shallow trench isolation (STI) process. It uses CMP to remove the excess deposited

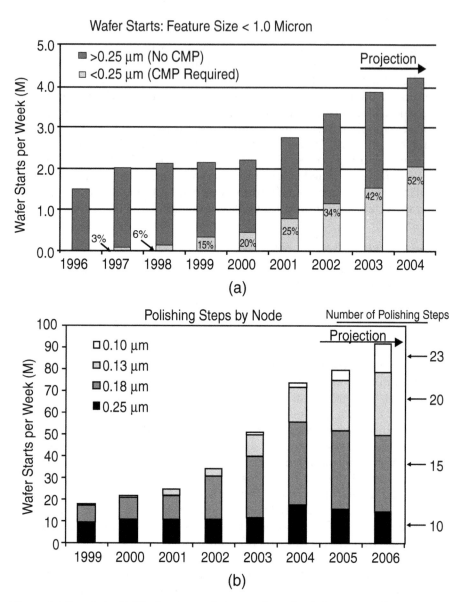

Figure 1.4 Trends in CMP adoption and the number of polishing steps.[16]

oxide. The two isolation processes are shown in Figure 1.6 (note that all of the active transistor elements are created after the isolation process is complete).

Although there are many other manufacturing steps involved in creating a CMOS circuit, the primary use of CMP involves the circuit interconnection lines that are fabricated to get the supply voltages or signals from one transistor to another.

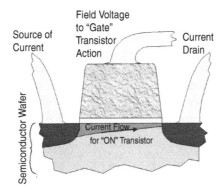

Figure 1.5 Basic field-effect transistor.

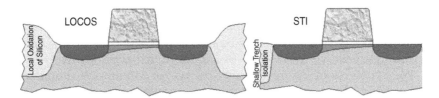

Figure 1.6 Methods of transistor isolation.

Early difficulties with interconnect lines involved issues of these metal lines going over steep steps, which induced reliability and electromigration problems. The method of contacting the interconnect metal to a previous layer involved etching a contact hole and depositing the metal so that it would dive down into the hole and return to its layer as the metal line traversed the surface (Figure 1.7). High topography contacts increased the probability of reliability problems (such as microcracking). Advanced circuit design companies realized that different contact methods would be needed to support more complex chips. Modern solutions involve CMP to remove excessive variation in surface height (topography).

As circuits get more complex, we increasingly need to remove the effects of high topography. Routing challenges increase very rapidly with increasing numbers of transistors. The simplest solution is to stack extra layers on top of each other to increase the interconnect routes available to the circuit. Each layer adds its own topography problems, compounded by the surface variations existing from the layers below. Within four or five layers, the problems become insurmountable, and planarization of the surface is required for reliability and yield.

Shrinking transistor size demands pattern definition tools that can print increasingly smaller features. However, increasing resolution for smaller features is obtained at the expense of the depth of focus. Depth of focus is the vertical range over which the image will be in acceptable focus (i.e., the image will adequately print). If the mask feature cannot be adequately imaged on

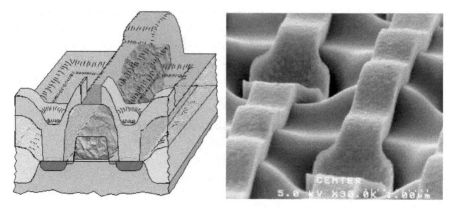

Figure 1.7 Early interconnect showing high topography contacts.

the wafer surface, the image size will be distorted, and some of the circuit elements will have positions and spacings different from their designed values. It will be up to the tolerances built into the circuit design whether the transistor element will work at all, or with degraded performance.

We can see from Figure 1.8b that the use of 248 nm UV illumination (with a 0.6 NA lithography tool) can yield a printable line of 0.25 μm resolution. However, it will provide adequate focus only over a range of about 450 nm. Since we need to place our focal point in the middle of the range, it leaves us with a little less than a quarter-micron variation in surface topography before we start losing control over our feature definition. Planar processing consists of adding layers of material on top of the existing surfaces and then subtracting the material in areas where we do not want it. Thus this process, which is the foundation of modern semiconductor processing, inherently creates steps in the surface topography each time a new layer is introduced. Many of the circuit elements are current-carrying conductors, and as such they need to have sufficient cross-sectional area to transport the current efficiently around the circuit. As the surface dimension features shrink for greater circuit compaction, the only dimension left to support a sufficient cross-section is the vertical height of the current-carrying conductors. This means that while the circuit size shrinks from generation to generation, the height of the vertical steps continues to be significant. The combination of these effects demands some form of planarization to reduce the circuit topography within the required depth of focus for accurate circuit imaging.

Two of the primary drivers for integrated circuit marketing success are cost and speed. Both elements happen to be correlated with the size of the circuit and its elemental components. Various physical defect mechanisms generate a distributed probability of occurrence. The smaller the integrated circuit chip (the die), the greater the chance of producing some working circuit chips with a given defect density. This was the initial driver for shrinking the circuit element dimensions (the size of the transistors, lines

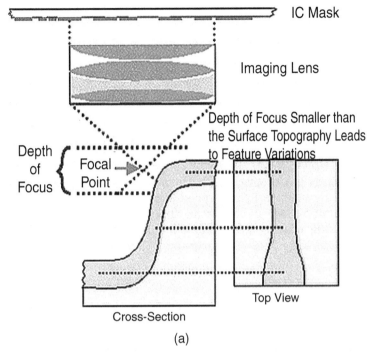

(a)

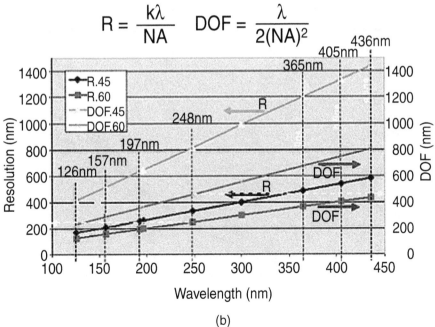

$$R = \frac{k\lambda}{NA} \quad DOF = \frac{\lambda}{2(NA)^2}$$

(b)

Figure 1.8 (a) Depth of focus printing problems. (b) Feature resolution (R) and DOF versus wavelength.

and contacts, etc.). Two fortuitous circumstances came to favor the shrinking approach: (a) smaller transistors actually switch faster, and (b) smaller circuits use less power (generating less heat, etc.).

1.2.4 Circuit speed issues

Integrated circuit speed relies on two primary determinations: (a) how long it takes the transistor to switch on or off, and (b) how long it takes the transistor to communicate its new state to the next transistor in the logical sequence. As the transistor gets smaller, the transistor switching speed increases. However, the speed of communication from one transistor to another tends to decrease as the circuit dimensions get smaller. At some point the transistor switching speed increases so much that it is no longer the rate-limiting step of the circuit speed. From there forward in the circuit-shrinking process, the struggle for greater speed centers around the interconnect delay between transistors (Figure 1.9). It is true that as the

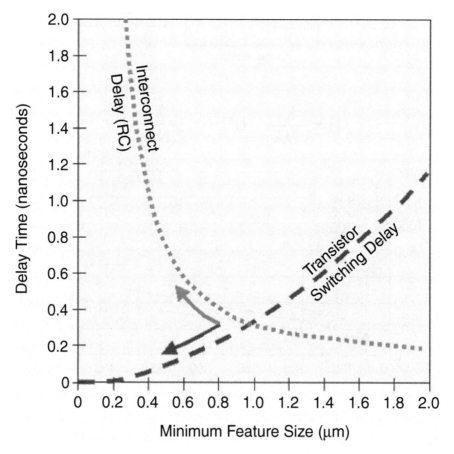

Figure 1.9 Basic integrated circuit delay components. (Based on Yang.[16])

Factors Effecting Capacitive Delay:
$$C = k\, a_1/d_1$$

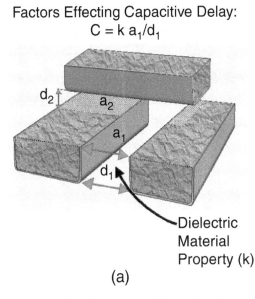

Dielectric
Material
Property (k)

(a)

Low-k Dielectrics

Material	k Value
SiO₂	3.9–6.0
F-SiO₂	3.2–3.6
C-SiO₂	2.6–3.2
Polyimides	2.9–3.9
Silesquioxanes	2.7–3.0
Polyarylethers	2.7–2.9
Fluorinated Polyimides	2.3–2.8
Parylenes	2.2–2.6
Porous Amprophous C	2.1–2.5
Fluoro-Polymers	1.8–2.2
PTFE	~ 1.9
Xerogel / Aerogel	1.8–2.2
Micro-Porous Polymers	1.7–1.3

(b)

Figure 1.10 (a) Capacitive delays from interconnect geometry and (b) dielectric material *k*-values.[19]

spacing between transistors gets smaller, the dimensions over which the signal needs to propagate from one transistor to another decrease. But then typically the circuit size has also expanded, increasing the average interconnect distance. But by far the toughest part of the interconnect delay is the combination of resistance and capacitance that the interconnect lines possess in their material and physical relationships. This signal speed reduction is known as RC delay because the delay itself turns out to be the product of the interconnect line resistance and the capacitance it experiences with the other surfaces around it. A diagram of the delay geometry is shown in Figure 1.10a. This delay is a function of the overlapping area (Figure 1.10a), the separation between the surface distance (d), and a factor proportional to the ability of the intervening material to support an electric field. Part of the capacitive delay comes from the design geometry of the circuit, and the other part comes from the choice of materials used to fabricate the circuit (in this case, the insulating dielectric).

As the circuit features shrink, the interconnect delays start to get very large very fast. Some manufacturers have asserted that production of circuits faster than 600 MHz could not be achieved with conventional metals and dielectrics. Since circuit speed is one of the fundamental drivers of circuit market success, these barriers had to be surmounted. The resistance of an interconnect line, like the capacitance, is derived from the design geometry of the circuit and a choice of the material composition of the line. Of the three dimensions (Figure 1.11a), the layout design of a circuit will determine how far a transistor is from the next circuit element with which it needs to

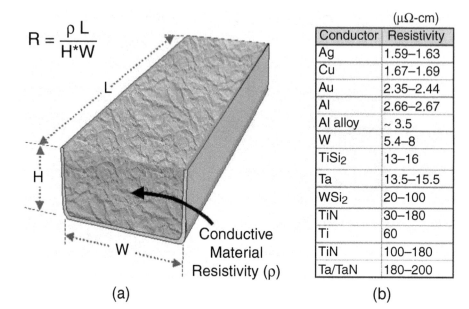

$$R = \frac{\rho L}{H*W}$$

	(μΩ-cm)
Conductor	Resistivity
Ag	1.59–1.63
Cu	1.67–1.69
Au	2.35–2.44
Al	2.66–2.67
Al alloy	~ 3.5
W	5.4–8
TiSi$_2$	13–16
Ta	13.5–15.5
WSi$_2$	20–100
TiN	30–180
Ti	60
TiN	100–180
Ta/TaN	180–200

Conductive Material Resistivity (ρ)

(a)   (b)

Figure 1.11 (a) Conductor geometry and (b) material resistivities.[18]

communicate (*L*). The other two dimensions (*H* and *W*) determine the cross-sectional area of the interconnect line. The resistance of the line is proportional to the length of the line, and inversely proportionally to its cross-sectional area ($R = \rho L/(W*H)$). The proportionality constant (ρ) is called *resistivity*, and is a property of the material chosen for the line. Most of the geometrical factors are not independently available to help the designer reduce the overall resistance of the circuit; as the circuit shrinks, the length *L* between transistors will be reduced (which decreases the resistance), but the thickness of the lines *W* will also be reduced (and a shrinking cross-sectional area increases the resistance). Because of these constraints, various efforts have been made to find new materials with lower resistivity that have properties suitable for manufacturing. The chosen interconnect material for the current leading-edge device generation is copper, which cuts the resistivity about in half. The principal drawback for normal manufacturing is that copper has no effective etch process. Semiconductor manufacturing is predominantly a subtractive methodology, where a layer is deposited on the surface and then etched away where it is not wanted. Without the ability to etch copper, normal processing could not take place. This is where CMP and the Damascene process come in. A trench is cut in the surface oxide (Figure 1.12), and the copper is deposited everywhere including the trench. The CMP process is then used to remove the copper that sticks up above the oxide surface, leaving copper in the inlaid line.

The overall circuit delay will be a combination of the transistor switching delay (which gets smaller with smaller feature sizes), and the interconnect delay (which gets larger with smaller feature sizes). The choice of the

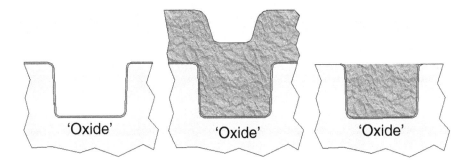

Figure 1.12 Copper Damascene process (cross-sectional view).

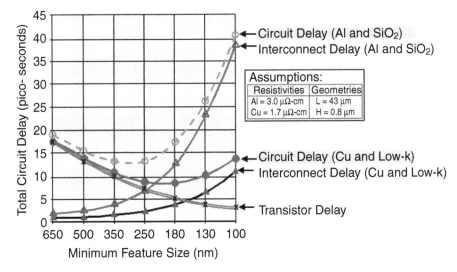

Figure 1.13 Integrated circuit delay trend for older and newer generation interconnects. (Based on ITRS National Roadmap.[21])

material to form the interconnect and the material to insulate between them has a big impact on the final circuit speed. Figure 1.13 illustrates the dependence of total circuit delay on these choices. (Note in Figure 1.13, the progress of shrinking feature sizes is positioned to show circuit evolution moving from left to right.)

1.2.5   Semiconductor road maps

The evolution of semiconductor manufacturing capability has always encountered tough and sometimes unexpected problems. As the R&D efforts demand ever larger budgets, greater cooperation among industry members is required to maintain progress. One such effort developed out of a joint program in 1988; leading scientific and academic institutions were brought together, and established a "road map" that could be used to forecast the difficult developments that must be overcome in order to continue providing the economic advantages of ICs. The results today are the various road maps that define the requirements of different technologies within the IC process (such as lithography, interconnects, etc.) and highlight the areas of concern, each with a forecasted time frame for their achievement.

The significance of the challenges for CMP is directly related to the ability of the industry to continue the progress of shrinking the IC device dimensions. Perhaps the best identification of these challenges and the industry's current abilities come to us in the form of an international assessment project called the ITRS (International Technology Roadmap for Semiconductors). Observation of the road map's challenges is an excellent way to identify the critical role that CMP plays in the health and future viability of the semiconductor industry.

1.2.5.1 An international road map for semiconductors

Although it was not readily apparent at the beginning of this project, the importance of international participation in defining and addressing the needs of the semiconductor industry is very clear today. The direction it takes is derived from the International Roadmap Committee (formed from industry associations in Japan, Europe, Korea, Taiwan, and the United States, which represent 95% of worldwide semiconductor production) and from the 12 international technical working groups (ITWG). Figure 1.14 shows the composition of the International Roadmap Committee.[20]

The forecast for semiconductor technology revolves around a key parameter called the *technology node*, which is used to track the critical size of the semiconductor device. The technology node is identified by two principal parameters: the technology node value and the technology node timing. Because of the diversity of IC devices based on their market segments, a single definition for the critical size of the semiconductor device is not possible. The ITRS tracks two device sizes in order to provide meaningful numbers across the spectrum of industry products: the devices of DRAM (dynamic random access memory) and MPU/ASIC (microprocessor unit application in specific integrated circuits). For DRAMs, the critical dimension is the half-pitch of the metal interconnects, whereas the MPU/ASIC critical dimension is the smallest half-pitch of the poly-wiring layer (see Figure 1.15[21]).

The technology node timing gives similar problems of definition, since a breakthrough device can appear in an R&D environment far in advance of its capability of being rendered in a pilot production, let alone a full production mode. The early definition of when a technology node had become available was "the year in which leading chip manufacturers begin shipping volume quantities (10,000/month) of product manufactured with qualified production tooling and processes."[21] A clarification was adopted in ITRS 2000, which added the constraint that this level of production by one

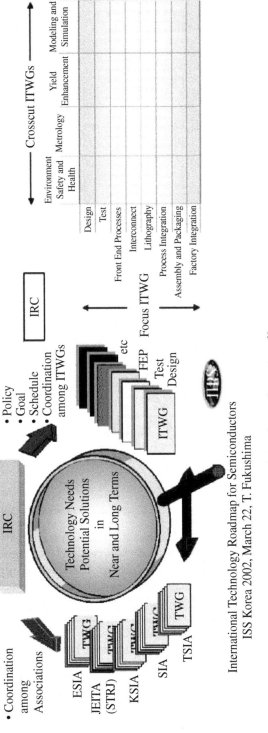

Figure 1.14 The composition of the International Roadmap Committee.[20]

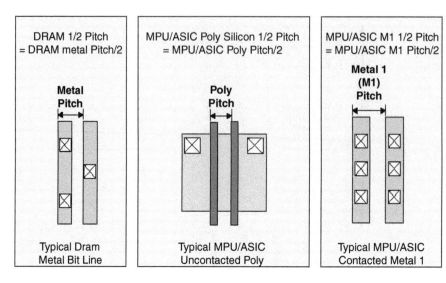

Figure 1.15 Illustration of minimum dimensions tracked by the ITRS Roadmap.[21]

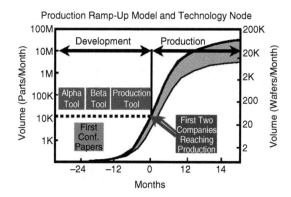

Figure 1.16 Typical production ramp curve illustrating technology node timing.[21]

manufacturer must be followed by another manufacturer within 3 months to establish the year that the technology node was realized (see Figure 1.16[21]).

One of the aspects of semiconductor technology that must be appreciated to get a proper perspective on its manufacturing demands is the scale within which it must work. Some common items are shown in Figure 1.17 along with the dimensional scales of the critical features of an IC[22] and a scanning electron micrograph of pollen grains on an IC.[23]

If we look at the projected scaling trends from the 2003 International Technology Roadmap (Figure 1.18), we can get a sense of the evolution of IC dimensional scales compared to the projected limitations of manufacturability. As circuit demands for planarity shrink down below 100 nm, the

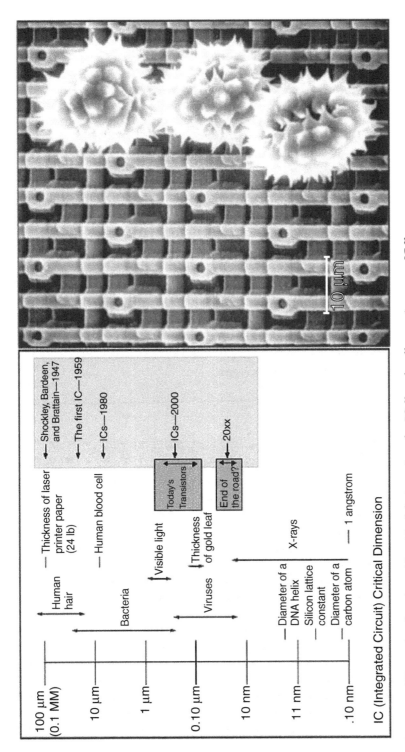

Figure 1.17 Dimensional scales with critical feature sizes for ICs[22] and pollen grains on an IC.[23]

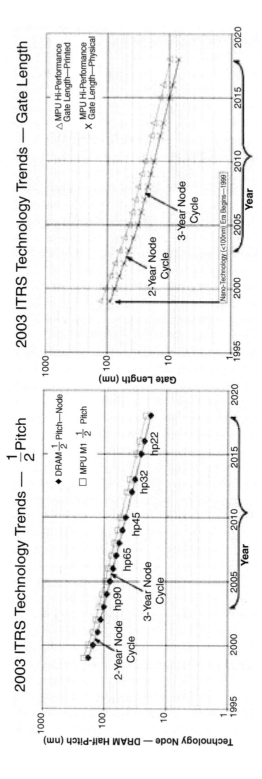

Figure 1.18 ITRS technology trends.

challenges for both processing and metrology become severe in a technology as complex as CMP.

Figure 1.19 shows the evolution of critical sizes in transistor dimensions. The line widths for various production parts are shown in Figure 1.19a,[24] while the research (as measured by journal publications) is shown in 1.19b.[25]

One interesting question is why electronics should be unique. In nearly every other product area, each year sees higher prices and shrinking quality or feature sets. But in electronics, it seems that frequently the price of the major commodity items goes down, and often with additional features included. The flagship product in this area is the personal computer. Although it is somewhat difficult to identify a metric with which to measure progress, one metric of productivity for comparison is processing speed per dollar of purchasing price. Figure 1.20 tracks this productivity for computers against time from 1900 to 1998. As an aside, it also positions the brain power equivalent level of computing power to provide a perspective (and a sense of humility).[26]

As for the uniqueness of electronics, we could look at things like increased manufacturing effectiveness, larger facilities (buying volume), or movement of manufacturing to areas of cheaper labor. But these would position this long-term trend of more than 30 years (see Figure 1.21[27]) as a pyramid scheme, since only by continually moving to larger factories and ever-cheaper labor could the trend continue.

This uniqueness lies in the ability of the semiconductor manufacturers to increase productivity (decrease of cost per function) year after year. There are many complex factors that contribute to manufacturing productivity (Figure 1.22[24]), but one of the key enablers is the continual decrease in the size of the transistor and other elements that make up the IC.

1.2.5.2 Moore's law

A comment[28] made by Dr. Gordon Moore in 1965, when he was a director of research and development at Fairchild Semiconductor, has evolved into the famous Moore's law. He observed at the time that the development of the IC was happening at a pace that provided twice as many transistors on an IC chip each successive year. He later adjusted the pace to about 60%, which predicts a doubling of the transistors on an IC about every year and a half.

"This trend, described by Moore as a 'simple observation,' has no necessary grounding in either economics or physics. … It was an expression of optimism, based on the technology available at the time."[29]

The interesting thing about the phenomenon that has grown up around that initial comment is that it is no longer considered a descriptor, but rather more of a determiner. "It is one of the 'laws' that underpin the strategic assumptions of virtually all the core information technology companies."[29] The acceptance of this trend has evolved into a defining strategy that continues to drive the semiconductor and information technology industries. The actualization of this "law" can be seen in Figure 1.23.

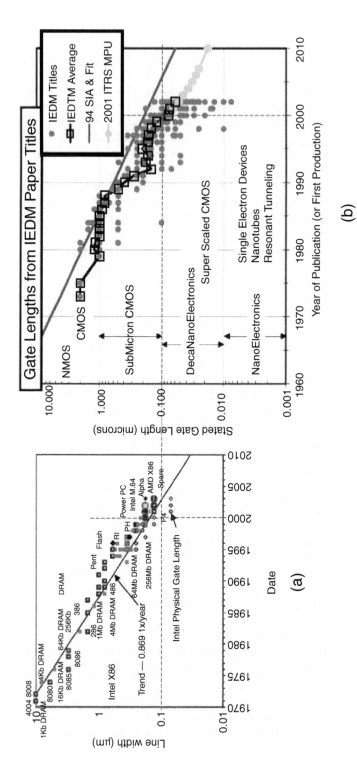

Figure 1.19 (a) Decreasing line width feature size for IC circuits[24] and (b) research work on shrinking transistor size.[25]

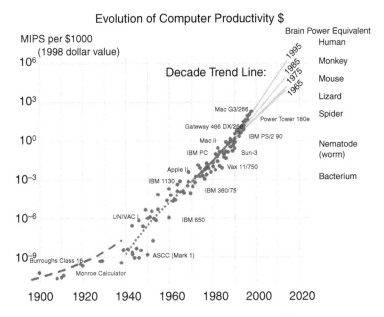

Figure 1.20 Computer productivity metric evolution (with brain power comparison).[26]

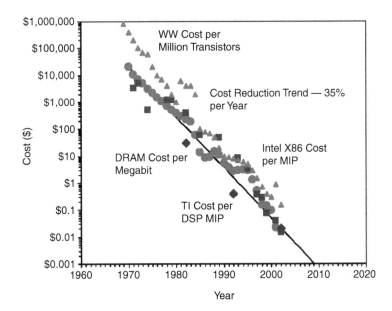

Figure 1.21 Product pricing trend. (From Jones.[27])

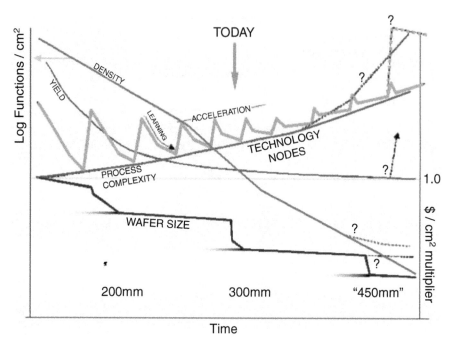

Figure 1.22 Productivity contributors and their form.[24]

Technological advance has now a frantic pace that leaves a wake of obsolescence behind it. When should we continue with the diminishing percentage profits from a mature product, and when should we take the larger percentage profits available from the introduction of more advanced products? Making such a decision is as much of an art as it is a science. The economics of semiconductor manufacturing is a complex balance of IC design, manufacturing, and marketing. The time it takes to bring a product to market plays a big role in the final profitability that a company may realize. The computation of the revenue and cost penalties of any particular commercialization date for a product (Figure 1.24[30]) can be difficult to gauge correctly.

Designing lithographic tools that can print very small images is quite complex, and the complexities increase very rapidly with smaller dimensions. The field size of this projected image has shrunk dramatically from the size of an entire wafer to the size of only a couple of circuits. This has necessitated a technology that steps and repeats the image across the wafer, but with a precision significantly smaller than the minimum circuit feature size. The complexity of these tools drives up costs that increase the investment needed to build leading-edge manufacturing facilities (Figure 1.25[31]).

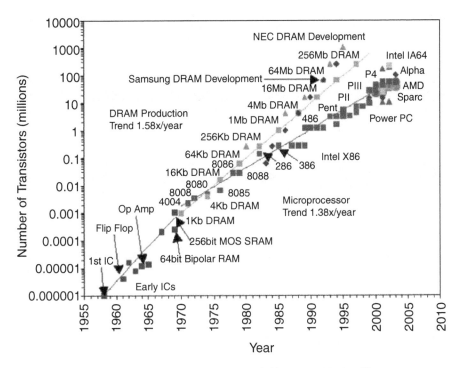

Figure 1.23 Trend in number of transistors available on an IC chip.[27]

1.3 The basic CMP process

A thorough discussion of the CMP process is not within the objectives of this book, but a good overview of the process steps and a discussion of some of the implementations can provide a useful foundation for understanding the tribology of CMP. Such an overview will be undertaken in this section. Although the authors try to avoid language that might be interpreted as commercial endorsements, one cannot review the CMP process without mentioning the equipment and support materials that are at the heart of CMP implementation.

An early understanding of material removal rate (MRR) from two surfaces in contact was developed by the glass polishing industry, where MRR became represented in what is known as Preston's equation: MRR = kPv. In this empirical equation, the MRR is seen as a linear function of the pressure between the working and the worked surfaces (P) and the shear velocity between the two surfaces (v). This equation has stood the test of time and still forms the basis of approaches to and control of the CMP process. However, it can readily be shown that neither P nor v can be primary variables in the removal of material from a surface. Material is only accelerated (removed) in the direction of an applied force. The interface pressure P represents a force per unit area but is applied compressionally and normal

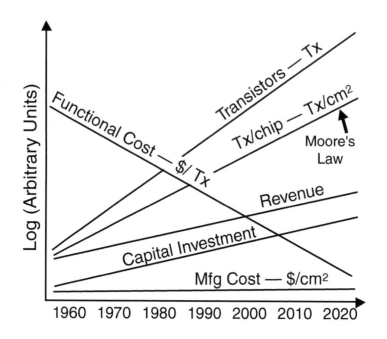

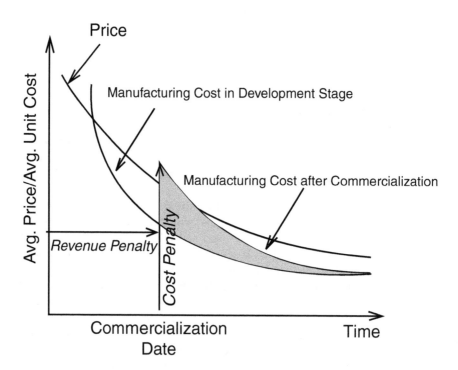

Figure 1.24 Semiconductor business trends[24] and manufacturing trade-offs.[30]

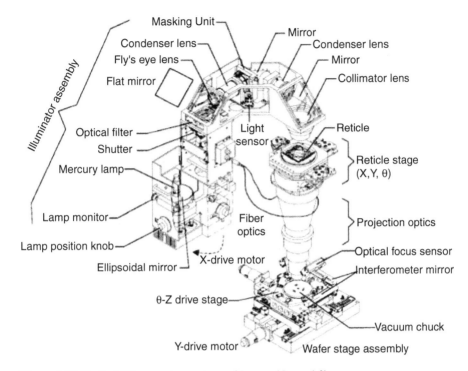

Figure 1.25 Optical lithography system of lenses (Canon).[31]

to the interface. With the exception of balanced momentum displacement, this force can only *add* material to the surface, not remove it. The velocity component is more promising in that it is applied in a shear direction, ideal for the direction of material removal. The velocity, however, is not a force and cannot by itself accelerate any material from the surface. Thus a deeper understanding of the way that these secondary variables couple with the viscoelastic pad material, slurry fluid momentum, and mobile abrasive impact must be reached before we can understand and predict the behavior of the CMP process at a fundamental level.

That said, however, we will use the Preston equation as a starting point in our discussion of the CMP process and the workings of the tools and materials used in its implementation. One of the reasons for using the simple variables of the Preston equation is that these are the variables most readily controlled by the tools of the technology. Thus the primary function of CMP tools is to deliver controlled downforce pressure and interface velocity to the surface of the wafer being polished (Figure 1.26).

The balance of forces as the wafer on the head engages the pliable pad material is critical to creating a planar surface. To compensate for the torque exerted on the head as the pad comes under initial compression and frictionally engages the leading edge of the wafer, the plane of the wafer is placed on a gimbal pivot. The downforce pressure from the shaft on the head

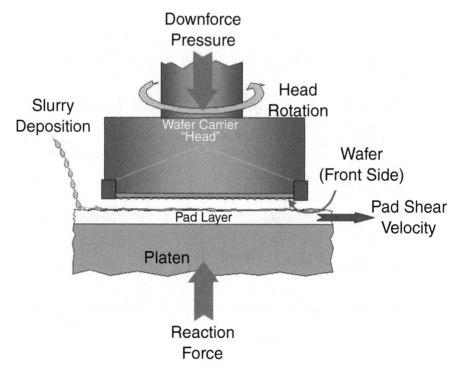

Figure 1.26 Basic CMP wafer head configuration.

is applied at the pivot point and then uniformly distributed across the wafer through the head design.

When several different material layers are involved, multiple planarization modules are used, and each module supports a different recipe. CMP tools provide tight feedback control over the principal Prestonian process variables, so recipes can be executed that sequentially vary parameters such as downforce pressure and pad velocity. But planarization of different materials often involves changes in the slurry-and-pad system to achieve the required performance. During processing, the pad is also treated with mechanical "stimulation," called conditioning, to maintain its performance properties. Thus multiple modules, each with its own slurry-pad- conditioning combination, are normally required to handle multimaterial planarization such as found in copper interconnect technology. When the head moves from module to module, the changing slurry involves a possible change in abrasive interstation.

Often several different robotic transport systems are used to convey the wafers to and from locations inside the CMP tool. Most tools have two basic internal environments under which they must operate: (1) a wet environment of multiple spray rinses, moist atmospheres, and high particle (slurry abrasive) conditions, and (2) a dry environment with an ultraclean low-particle

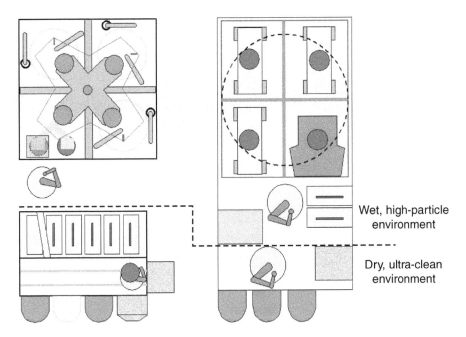

Figure 1.27 Rotary and linear CMP tools showing multiple module and multiple environment architecture.

atmosphere. Typically different robots service each environment to avoid cross-contamination (Figure 1.27).

One of the realities of the manufacturing process is that successive deposition layers forming the IC are often not of uniform thickness from the center of the wafer to the edge. CMP tools have evolved in a variety of ways to modify their removal rate locally in an effort to compensate for these incoming material thickness variations. The design evolution of rotary, linear, and orbital CMP tools has progressed to the point where they have become differentiated from each other (drastically in some aspects). As a result, the implementation of some features, such as localized pressure modification, operates through entirely different mechanisms and presents different capabilities and limitations (Figure 1.28).[32,33]

Another aspect of equipment process control relates to temperature. The mechanical component of CMP is based on friction and thus is a thermal energy source. The understanding of friction under shear force conditions is central to CMP performance. The subject is covered in Chapter 3 on friction. Often different materials in the planarization process have their own energy-balance reactions, and some are exothermic in their own right. Since there is always some form of chemical interaction to complement the mechanical actions in CMP, temperature will have an effect on the process rate. Some processes, like copper planarization, are heavily chemical in nature, and thus control variations are sensitive to temperature fluctuations.

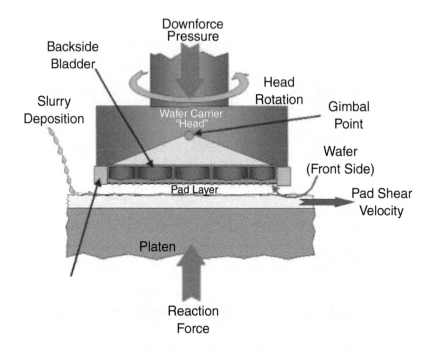

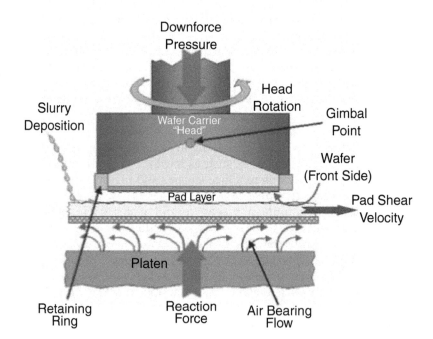

Figure 1.28 Details for rotary and linear wafer carrier heads with wafer pressure adjustments.

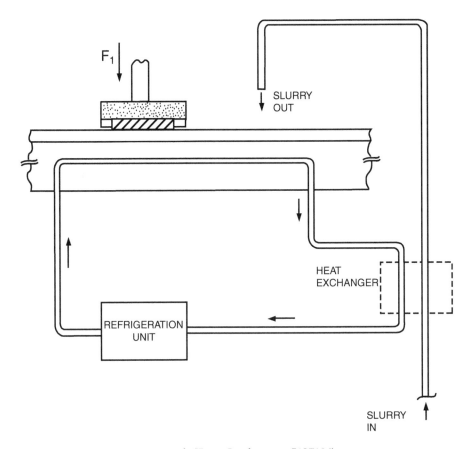

Figure 1.29 Temperature control. (From Intel patent 5127196).

Various tool architectures have addressed the problem in different ways. Fixed platen architectures often control the temperature of the platen through a recirculated heat exchange method (Figure 1.29), and sometimes they also control the temperature of the slurry being transported to the pad surface.

1.3.1 Scales

Surface texture and roughness play central roles in the theory of tribology, and we will explore them in the surface characterization section. One of the interesting elements that emerge is that the notion of roughness is interwoven with the perspective of scale (both in regard to physical dimensions and time). So it will be of use to review the scales that are encountered in CMP technology.[34]

As we shall see in the surface characterization and force transmission chapters, both the physical dimensions and the dimension of time over which the CMP mechanisms act are important in giving a proper perspective to the critical processes. Many components and subcomponents are

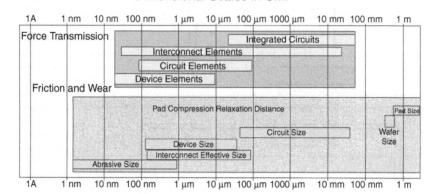

Figure 1.30 Physical dimensional interactions during CMP.

at work during the CMP process, each of which can contribute its own distinct signature in both the physical and the time-dimensional scale. Figures 1.30 and 1.31 provide a pair of rough illustrations. The extensive range that is encompassed by elements in the CMP process (nearly nine orders of magnitude in size and nine orders of magnitude in time — if equipment stability issues are left out) is the first thing to become apparent. Although no one element spans the entire breadth of these dimensions, each element is important and contributes uniquely to the final performance of the technology.

Thus mechanisms in each of these dimensional regimes must be appreciated in their own right, as well as in the context of their interrelations. This need to appreciate interaction mechanisms at each dimensional level provides distinctive challenges to those who attempt to create a unified model of CMP technology performance. Today's sophisticated models are needed both to characterize and to bridge these dimensional challenges.[35]

The basic process of planarizing a surface in CMP consists of compressing the wafer against an abrasive surface and applying shear force to create regions of high friction and wear.

The wafer is often mated against a pliable polymer material referred to as the pad. The abrasive nature of this mating surface is established by coating it with a mixture of chemicals carrying minute particles (the combination of chemicals and abrasive particles is called a *slurry*). Two substantial variants of this scenario are (1) a process in which the pliable pad is replaced by a predefined structure of abrasive islands (in what is called a *fixed abrasive* process), and (2) a process in which the pad polymer itself is used as the abrasive wear surface (called the *abrasive-free* process). These variants are driven by cost and special performance requirements, but they are not currently considered to be in the mainstream of CMP manufacturing. All this presents a rather generalized view, and much greater complexity is needed by the time a true manufacturing process is realized.

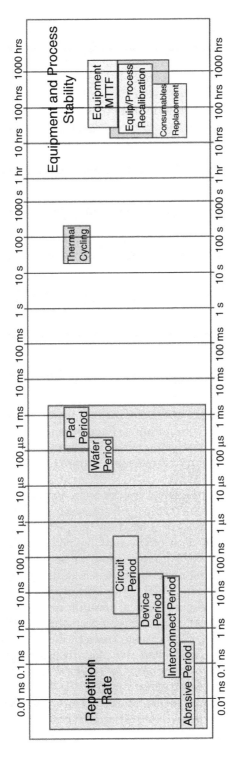

Figure 1.31 Temporal dimensional interactions during CMP.

The primary metrics of success for a CMP process are (1) the MRR and (2) the post-CMP defect density. But even the removal rate is made up of a complex of other factors, which have mechanisms operating from the nanoscale to the macroscale. A nanotribology view of the process would observe the individual abrasive particles in their interactions, sandwiched between the pliable pad and the exposed wafer surface. If we were to zoom out our viewpoint one order of magnitude or two, we would encounter the variations in wafer surface features (both physical dimensions and material variations) that constitute the elements of the IC device components, such as interconnect lines.

Some elements are packed closely together, while others are separated by vast distances (in our dimensional scale). The ability of our abrasive-rich pad surface to apply a uniform force to those surface features varies with their spacing (this is called the *pattern density* effect). If we zoom out another few orders of magnitude, we now see very substantial differences in the surface uniformity of the polishing pad, and our viewing distance can now span across a whole IC (die). We need to be removing material uniformly across the circuit so that all parts of the IC will function correctly and we can reasonably pattern the next layer step in the process. Zooming out another couple of orders of magnitude will encompass the entire silicon wafer. For best product yield, we need to be removing material uniformly across all circuits on the wafer. But at this macro level, the boundary forces compressing the limited-radius wafer into the larger pliable pad create some serious nonuniformities of the force density distribution. Also at this level, the even coating of slurry suffers uniformity problems as the pad surface encounters the leading edge of the compressed wafer. Many variants of the hardware and processing configurations have been developed to reduce the severity of these effects as much as possible. These variations are reflected in the increased subset of removal rate metrics used to judge the performance of the CMP process.

The circular shape of the wafer invites a circular coordinate system for tracking variations (angle and radius). To smooth out local variations in the removal process, the wafers are rotated on the pad about their centers so that the approaching direction of the moving pad surface will gradually impinge on the wafer from all sides (more or less uniformly). Thus the primary variable describing nonuniformities is radial (center-to-edge).

A smaller scale metric for MRR nonuniformity involves examining the wafer surface material thickness (vertical nonuniformity), which results from force variants. Two different types of effects occur, both of which result from variation in surface material or spacing (pattern density). The first type of pattern density effect occurs when the planarization acts on a uniform material that has nonuniform topography (vertical steps). The metric used for this is often called *the planarization length* (L) or *the planarization efficiency* (Figure 1.32). The second type of pattern density effect is driven by the horizontal surface dimensions among different types of materials (which can

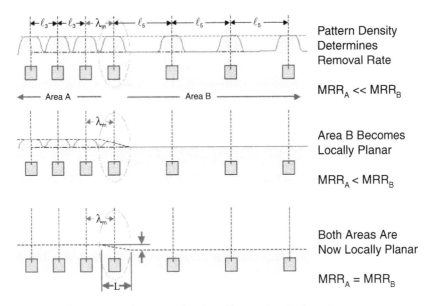

Figure 1.32 Illustration of pattern density effect on local planarity.

be thought of as nonuniform material even on uniform topography). The metrics of this type are called *dishing* and *erosion*.

The primary function of CMP is to planarize the surface of the wafer through the removal of surface material. A working definition of planarization is: to make planar or flat by removing elevated topographical features. But this does not capture the material removal aspect, which can be nearly as important as making planar, for such applications as copper interconnect technology.

"Key knowledge of the chemical, structural, and mechanical properties of the surface to be polished establishes the polishing parameter space ...."[36] Such knowledge must also include the equivalent list of properties of the mating surface that comes in contact with the wafer surface under polish.

The pad surface structure, the type and size of the abrasive particles, and the chemical balance of the slurry must be tailored to the materials being planarized. But the results will also depend on polish tool parameters such as downforce, pad speed, slurry flow rate, and processing temperature. In addition, the effects can change dramatically in some cases depending on the structural spacing and step height exhibited by the circuit devices being planarized. There will be discussion throughout this book of the pad–slurry system, but in truth it is a wafer–pad–slurry system that sets the polishing parameter space. And even then, the term *wafer surface* refers not only to the primary material being presented to the pad, but also to any other material as well as the relative vertical and horizontal spacing (topography).

A brief discussion of the input and output variables in CMP will allow for a more integrated overview of the process. If one were to try to sum up the CMP process in a single word, it would probably be "interdependencies." The Ishikawa illustration (Figure 1.33[37]) is a useful way of identifying the principal components of the CMP process and relating their dependencies. To the left of the center rectangle are the process parameters, and to the right are the performance parameters. The Holy Grail of process is the mapping of these process parameters to the performance results. This mapping can be explored empirically or mechanistically. It is widely acknowledged, however, that many of the significant physical and chemical mechanisms of CMP are not fully understood.[15] Thus much of the primary work is done as an empirical mapping through the design of experiments.

The complexities of the process are enhanced by the variety of choices for classifying the parameters involved. Figure 1.34 shows five different approaches that could be used to partition the same set of process and tool parameters. A variation on the classic blind men describing an elephant (Figure 1.35) could be used to illustrate how the choice of the aspect paradigm changes the vocabulary used for parameter partitioning of CMP.

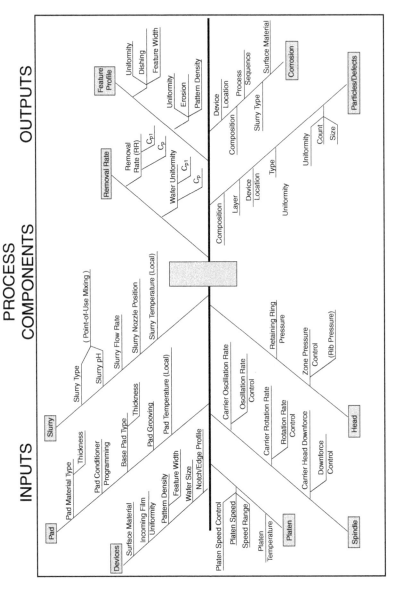

Figure 1.33 Example input and output process components diagram for typical CMP configuration.

Parameter Partition

Temporal Aspects Fundamental Interface Aspects Environmental Aspects

Static Parameters   Force Transmission   Physical Environment

Dynamic Parameters Friction and Wear   Chemical Environment

         Fluid and Surface Chemistry   Temperature Environment
                     (Interfacial Friction)

         Manufacturing Aspects   Process Input Aspects

         Tool Basic Parameters

         Process Execution Parameters ⟷ Recipe

         Consumables ⟵⟶ Material Parameters

         IC Parameters ⟶ Pattern Density Parameters

Figure 1.34 Different approaches to parameter partitioning of the CMP process.

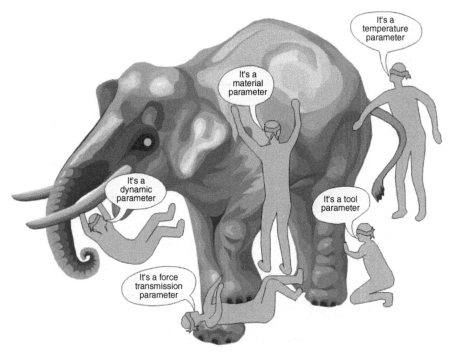

Figure 1.35 Variation of the classic blind men describing an elephant for CMP.

References

1. P. Jost, Lubrication Tribology — A Report on the Present Position and Industry's Needs, Department of Education and Science, HMSO, London, 1966.
2. B. Bhushan, *Introduction to Tribology*, John Wiley, New York, 2002.
3. K.M. Ramanan, Neandertals: A Cyber Perspective, http://sapphire.indstate.edu/~ramanank/.
4. D. Dowson, *History of Tribology*, 2nd ed., Institution of Mechanical Engineers, London, 1979.
5. J.A. Williams, *Engineering Tribology*, Oxford Science Publications, 1994.
6. D. Dowson, *History of Tribology*, Professional Engineering Publishing, 1998.
7. Leonardo da Vinci Drawings: History of Science Friction. http://www.tribologie.nl/backgrounds/history/history.htm.
8. G. Amontons, De la résistance causée dans les machines, *Mémoires de l'Académie Royale*, A, 257–282, 1699.
9. C.A. Coulomb, Théorie des machines simples, en ayant regard au frottement de leurs parties, et à la roideur des cordages, *Mém. Math. Phys.* (Paris), 10, 161–342, 1785.
10. W.F. Parish, Three thousand years of progress in the development of machinery and lubricants for the hand grafts, *Mill and Factory*, 16 and 17, 1935.
11. C.M. Mate, Application of disjoining and capillary pressure to liquid lubricant films in magnetic recording, *J. Appl. Phys.*, 72, 3084–3090, 1992.
12. C.M. Mate, Atomic-force-microscope study of polymer lubricants on silicon surface, *Phys. Rev. Lett.*, 68, 3323–3326, 1992.
13. C.M. Mate, Nanotribology of lubricated and unlubricated carbon overcoats on magnetic disks studied by friction force microscopy, *Surf. Coat. Technol.*, 62, 373-379, 1993.
14. B. Bhushan, J.N. Israelachvili, and U. Landman, Nanotribology: friction, wear and lubrication at the atomic scale, *Nature*, 374, 607–616, 1995.
15. M.A. Fury, The early day of CMP, *Solid State Technology*, May 1997.
16. G. Yang, Outlook for CMP Consumables, Salomon Smith Barney, 2002.
17. S.P. Jeng, R.H. Havemann, and M.C. Chang, *Mat. Res. Soc. Symp. Proc.* 337, 25, 1994.
18. http://www.goodfellow.com, and *Handbook of Chemistry and Physics*, 85th ed., CRC Press, Boca Raton, 2004.
19. R. Zorich, *Handbook of Quality Integrated Circuit Manufacturing*, Academic Press, Washington, DC, 1991.
20. T. Fukushima, An Overview of the ITRS 2001, ITRS: http://public.itrs.net.
21. ITRS, International Technology Roadmap for Semiconductors, 2003, http://public.itrs.net/Files/2003ITRS/Home2003.htm.
22. A. Penzias, The next fifty years: Some likely impacts of solid state technology, *Bell Lab. Tech. J.*, 2, 155–168, 1997.
23. H. Föll, Electric materials, Lecture note, Faculty of Engineering, University of Kiel, 2004.
24. R. Goodall, D. Fandell, A. Allan, P. Landler, and H.R. Huff, Long-term productivity mechanisms of the semiconductor industry, *American Electrochemical Society Semiconductor Silicon 2002 Proceedings*, 125, 2002.
25. Nishi_GateLengthEvolution1.gif.
26. H. Moravec, When will computer hardware match the human brain?, *J. Evol. Technol.*, 1, 1998; http://www.transhumanist.com/volume1/moravec. htm.

27. S.W. Jones, Exponential Trends in the Integrated Circuit Industry, http:// www.icknowledge.com, 2003.
28. G. Moore, Cramming more components onto integrated circuits, *Electronics*, 1965.
29. Beyond Moore's law, Computer Business Review Online (www.cbronline. com), Jan. 1997.
30. N.W. Hatch, J.T. Macher, Mitigating the Tradeoff between Time-to-Market and Manufacturing Performance: Knowledge Management in Developing New Technologies, mimeo, Brigham Young University, 2003.
31. M. Quirk, J. Serda, *Semiconductor Manufacturing Technology*, Prentice Hall, New York, 2001.
32. T. Bibby and K. Holland, *XEquipment, Chemical Mechanical Polishing in Silicon Processing*, S.H. Li and R.O. Miller, Eds., Academic Press, 2000, chap. 2.
33. D. Wei, J.M. Boyd, Y. Gotkis, and R. Kistler, Edge Control for Removal Uniformity by Air Bearing Technology on a Linear CMP System, SEMICON Taiwan 2001, Taipei, September 17–19, 2001.
34. H. Dagnall, *Exploring Surface Texture*, Taylor Hobson PNEUMO, Leicester, U.K., 1980.
35. M.R. Oliver, CMP fundamentals and challenges, *Mat. Res. Soc. Symp. Proc.*, 566, 73, 2000.
36. J.M. Steigerwald, S.P. Murarka, and R.J. Gutmann, *Chemical Mechanical Planarization of Microelectronic Materials*, John Wiley, New York, 1997.
37. K. Ishikawa, *What Is Total Quality Control?*, Prentice-Hall, New Jersey, 1985.

chapter two

Surface properties

Real surfaces are covered with asperities. The texture of a surface is complicated, and many parameters are used to quantify the various surface characteristics. This practice is common in microelectronic industries. There are
three primary methods of numerically representing surface roughness: centerline average (R_c), root-mean-square (R_q), and maximum peak-to-valley
height (R_t). These are defined as:

1. Centerline average (CLA) is the most common designation of surface
   roughness and is also called the arithmetic average (R_a). Here R_a is
   calculated from:

$$R_a = \frac{1}{N} \sum_{i=1}^{N} |Z_i| \qquad (2.1)$$

   where N is the total number of measurements and Z is the absolute
   peak-to-valley height with respect to a reference line.
2. The root-mean-square (RMS) representation of a specific height (R_q)
   is:

$$R_q = \left[\frac{1}{N} \sum_{1=1}^{N} Z_i^2 \right]^{1/2} \qquad (2.2)$$

3. Maximum peak-to-valley. The maximum peak-to-valley asperity
   height (R_t) is:

$$R_t = R_{max} - R_{min} \qquad (2.3)$$

   Typically the magnitudes of these peak asperity height measurement
   methods follow the relative order:

$$R_a \leq R_q \leq R_t \qquad (2.4)$$

Surface roughness can also be represented by a surface roughness grade base on the R_a value.[1] Surface roughness is also referred to as R_a (average roughness) or it may be referred to as roughness grade.

Experimental surface roughness data, typically obtained by profilometry, is numerically analyzed[2] to obtain an average roughness value such as those provided in Table 2.1. During numerical analyses of the experimental data, it is necessary to establish a roughness line through the topographical peaks and valleys. Data analyses are usually performed, using a computer, by one of the three methods shown in Figure 2.1.[3]

1. M-System. A line is selected that passes through the topographical profile so that the areas above and below the line are equal, as shown in Figure 2.1a.
2. Ten-Point Average. A line is drawn through the center of the five highest peaks and the five lowest valleys, as shown in Figure 2.1b.
3. Least-Squares Reference. A least-squares average through the topographical peaks and valleys is determined, as shown in Figure 2.1c.

Units used for surface roughness measurement depend on the systems applied. Table 2.1 lists the conversions of common data.

The purpose of chemical-mechanical planarization (CMP) is to obtain an atomically smooth wafer surface. The average roughness R_a of a polished wafer is less than 4 Å. For such a smooth surface, an atomic force microscope (AFM) is generally used. Figure 2.2 is an example of a polished wafer showing a supersmooth surface. During CMP, the planarization is defined using the term within-wafer nonuniformity (WIWNU).

Table 2.1 Surface Roughness Units and Interconversions

Roughness Values (R_a)		Roughness Grade
μm	μin	
50	2000	N12
25	1000	N11
12.5	500	N10
8.3	250	N9
3.2	125	N8
1.6	63	N7
0.8	32	N6
0.4	16	N5
0.2	8	N4
0.1	4	N3
0.05	2	N2
0.025	1	N1

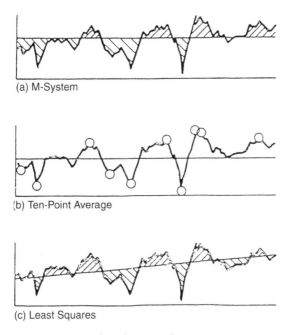

(a) M-System

(b) Ten-Point Average

(c) Least Squares

Figure 2.1 Numerical analysis of surface roughness.

Figure 2.2 Example of polished wafer showing supersmooth surface.

2.1 Surface conformity

To understand the polishing process, it is necessary to determine the load acting on the contacting surfaces. The geometry of surface contacts may be classified as either conformal or nonconformal (or counterformal).

Conformal surfaces are commonly used in journal sleeve bearings as well as hydrodynamic thrust bearings that fit together geometrically. During

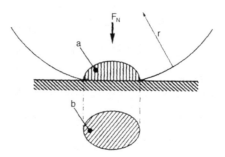

Figure 2.3 Hertzian contact.

CMP, a pad is generally soft and elastic; therefore the pad and wafer surfaces are correspondingly conformed. The loading of conformal surfaces is typically carried out over a wide area of peak asperities, and the load-bearing surface area remains nearly constant with increasing loads. The lubrication contact area is relatively large. Nonconformal surfaces are often found in two mating gear teeth, in vanes-on-rings, and rolling elements in their raceways, which do not fit together geometrically, and the load is concentrated on a relatively small contact area that increases with increasing load. The contacting area is called the *Hertzian* contact, as shown in Figure 2.3. Whether contact is conformal or nonconformal depends on the scale. A seemingly nonconformal pair of surfaces might be conformal at the atomic level. A pair of worn surfaces can also be conformal.

2.1.1 Real area of contact

When two solid objects are placed in contact, some regions on their surfaces will be close together, others further apart. It is known that atom-to-atom forces have very short ranges, so that all interaction takes place at those regions where there is atom-to-atom contact. These regions are referred to as junctions. The sum of the areas of all the junctions constitutes the real area of contact, A_r. The total interfacial area, consisting both of the real area of contact and of those regions that appear as if contact might have been made there, will be called the apparent area of contact, A_a.

The nature of the interaction between two surfaces is determined by the real area of contact, especially by the size of the real area. With a simple limit analysis assuming ideally plastic deformation, we can try to calculate a minimum value for A_r. If the surfaces that are placed in contact are rough, but not excessively rough, a typical junction will be induced as shown in Figure 2.4. Then the interface will be in a state of triaxial constraint. Thus the value of the real area of contact, A_r, is given by:

$$A_r \geq \frac{L}{p}$$

(2.5)

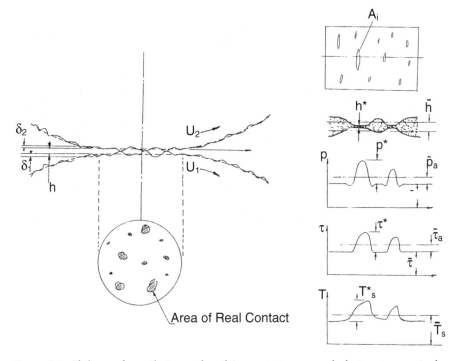

Figure 2.4 If the surfaces that are placed in contact are rough, but not excessively rough, a typical junction will be induced.

where L is the force normal to the interface of contact and p is the penetration hardness of the material.

When two surfaces with very small surface asperites are brought together, there may be no plastic deformation whatsoever, only elastic deformation, and A_r will be greater. That means $A_r > L/p$. This is true of a highly polished surface such as that of a bearing ball that is pressed into a flat polished surface. It will produce an area of contact as given by Hertz's equation for elastic deformation, namely:

$$A_r = 2.9 \left[Lr \left(\frac{1}{E1} + \frac{1}{E2} \right) \right]^{2/3} \tag{2.6}$$

One way to represent surface roughness effects on lubrication, particularly fatigue life failure, is to use the lambda (Λ) factor correlation [a].[2]

$$\Lambda = h_o / \sigma \tag{2.7}$$

where h_o is the lubricating film thickness and σ is the average surface roughness for the two surfaces coming into asperity contact; it is defined as:

$$\sigma = (\sigma_1^2 + \sigma_2^2)^{1/2} \qquad (2.8)$$

Although it is not completely understood, CMP undergoes wear in all modes in a conventional tribological process: abrasion, adhesion, electrochemical wear, and dissolution. In this chapter, we introduce basic wear concepts and discuss their mechanisms during CMP occurring on wafer and pad surfaces.

2.2   Surface properties

Here we discuss the properties of solids that are to be polished during CMP. Surface properties are important for CMP because they are related to the real area of contact, friction, wear, and tribochemical interactions between surfaces of wafers, slurry particles, and polishing pads. Upon contact, there are four structural elements involved in the wear mechanisms. They are, as shown in Figure 2.5,[4] surface films that are present < 1 μm from the surface, near-surface structure occurring between 1 and 150 μm from the surface, subsurface structure that occurs between 50 and 1000 μm from the surface,

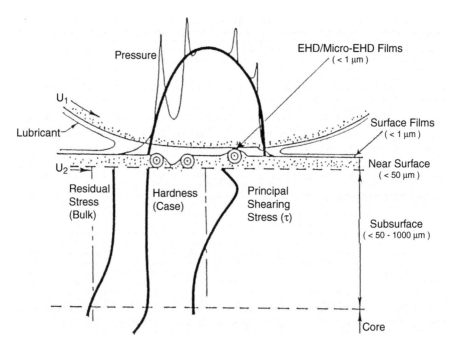

Figure 2.5 When surfaces are in contact, there are four structural elements involved in wear mechanisms.

and bulk material properties.[4] CMP is able to produce a surface roughness within 10 Å, one or two molecular layers. Any damaged areas larger than that are considered defects. One type of surface film is an oxide film that is formed by surface oxidation via oxygen present in the atmosphere, dissolved in the lubricant, or both.

The classic Barus equation describes the effects of pressure on viscosity:[5]

$$\eta_P = \eta_0 e^{\alpha P} \tag{2.9}$$

where η_P is the centipoise viscosity at pressure P, η_0 is the viscosity at atmospheric pressure, and α is the pressure–viscosity coefficient. Important refinements in the equation for the pressure–viscosity coefficient have been reported by Jones et al.[6,7] Since CMP uses water-based slurry, the pressure–viscosity coefficient should not affect CMP performance as much as oil-based lubricants do.

2.2.1 Surface films

In traditional tribological applications, many types of surface films are used to reduce wear. Here we provide a brief summary for reference. Physical or chemical adsorption is the second way of protecting film for lubrication. A thin surface film is formed by the adsorption of polar lubricant molecules onto the surface, providing an effective barrier against material-to-material contact. Temperature affects the effectiveness of the adsorbed film. A thicker film provides better protection of the surface. Two types, adsorbed film and reaction film, are shown in Figure 2.6. Physical adsorbed films, showing physisorption, are shown in Figure 2.7a. Chemically adsorbed films, showing chemisorption, are shown in Figure 2.7b.[8] Physisorption is a thermodynamically reversible, relatively weak adsorption process (no chemical bond formation) based on dipole or van der Waals forces. Chemisorption is thermodynamically irreversible and is based on considerably stronger ionic bond formation with the metallic oxide wear surface. Adsorption forces between a carboxylic acid and a metal substrate due to hydrogen bonding and Debye orientation forces have been reported in the range of 13 to 15 kcal/mol.[9] A hydrogen exchange interaction for fatty acids on a metal substrate such as copper has been postulated, as shown in Figure 2.8.[10] Metal surfaces are often modified by the formation of reaction films. Some reaction films are formed during heat treating processes such as carburizing, carbonitriding, and nitriding. Others are formed *in situ* by surface chemical reactions between an additive and the surface. Some surface films are formed by tribochemical processes. Various examples of tribopolymerization have been reported. For further information, please refer to References 11 through 14.

During CMP, the surface film, the passivation, is crucial for metal removal. We will discuss this in detail in the chapter on wear and removal mechanisms.

Surface Films

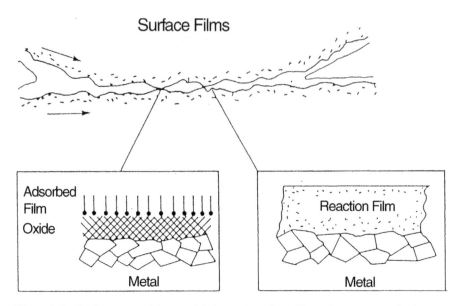

Figure 2.6 Surface asperities and lubrication thin films determine tribological behavior and types of lubrication modes.

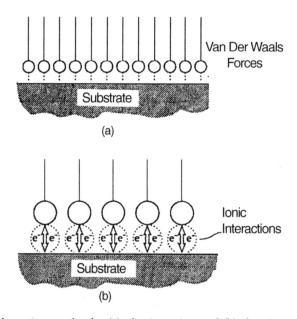

Figure 2.7 Adsorption modes for (a) physisorption and (b) chemisorption.

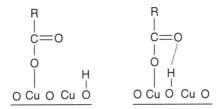

Figure 2.8 A hydrogen exchange interaction is shown for fatty acids on a copper substrate.

2.2.2 Near-surface structure

Near-surface structure occurs up to approximately 50 µm and is affected by surface hardening or reaction film formation in this regime. Near-surface structure is particularly susceptible to severe deformation and is important because the residual stresses in these regions will affect crack propagation during lubrication failure.

2.2.3 Subsurface structure

Subsurface structure predominates at depths of 50 to 1000 µm. This region is susceptible to gross deformation and is affected by the bulk properties of the material. Spalling failure typically occurs in this region.

2.2.4 Bulk material properties

Some of the most common bulk material properties that influence wear are: the melting point, Young's modulus, yield strength, hardness, and surface energy. Table 2.2 provides a summary of these values for selected pure metals. To provide optimal antiwear properties, the material should:

1. Possess high strength to resist plastic flow.
2. Have high ductility to withstand repeated plastic strain without cracking.
3. Be homogeneous and free from discontinuities, such as impurities and soft transformation products.

A smooth and homogeneous material generally has better antiwear properties because failure takes place at higher energy areas such as boundaries and rough sites.

Table 2.2 Properties of Metallic Elements[15]

	Melting Temp. (°C)	Young's Modulus (E) (dyn/cm²)	Yield Strength (σ_y) 10^9 (dyn/cm²)	Hardness (P) (K²/mm²)	Surface Energy (erg/cm²)
Aluminum	660	0.63	1.1	27	900
Cadmium	270	0.32	—	7	390
Copper	1083	1.2	3.2	80	1100
Magnesium	650	0.44	1.5	46	560
Manganese	1245	—	2.5	300	—
Nickel	2.08	—	3.2	210	1700
Tin	232	0.44	0.15	5.3	570
Zinc	420	0.91	1.3	38	790

2.3   Surface characterization techniques

2.3.1   The profilometer

Surface roughness or surface topography, if it is measured in two or three dimensions (2D, 3D), is determined experimentally using a profilometer. Profilometers are either contacting or noncontacting. Contacting profilometers use a wire, or a stylus, to contact the surface while tracing the surface asperities and shape. Noncontacting profilometers use either acoustical or light signals to map the surface asperities. This is probably the most widely used method of measuring surface roughness. The stylus usually has a rounded end made up of some very hard material. In a roughness measurement, the stylus is moved across the surface and its vertical movements are amplified electrically or mechanically and usually are recorded on a chart. Alternately, the small displacements are integrated and averaged to produce an arithmetic average R_a or a root-mean-square R_q roughness value, representing average departure of the measured profile from its ideal position.

The performance characteristics of a profilometer are determined by the shape and size of the probe (the radius of curvature of the stylus tip is typically 5 µm). Although profilemeters are extremely sensitive in detecting gentle undulations in a surface, they are quite poor at detecting sharp crevasses and give a highly distorted impression of sharp ridges. Their basic resolution in the vertical direction is usually of the order of a microinch.

2.3.2   Optical techniques

Although a profilometer can provide a quick and convenient indication of the surface roughness along one line, it is not convenient for assessing a

Table 2.3 Methods Used in the Measurement of Surface Roughness

Method	Magnification Range Lowest	Magnification Range Highest	Resolution at Maximum Magnification (Microns) Horizontal	Resolution at Maximum Magnification (Microns) Vertical	Depth of Field (Microns) Lowest	Depth of Field (Microns) Highest
Optical microscopy	×20	×1000	0.5	0.01	5	0.5
SEM	×20	×10^5	0.015	1	1000	0.2
TEM	×2000	×2×10^5	0.0005	0.0005	5	0.1
Surface profilometry	×50	×10^5	5	0.05	1	0.5

substantial area for departures from smoothness. For the latter purpose it is more convenient to resort to an optical technique. Generally, a beam of light is shone at the surface, and then we inspect to see how the beam has been disturbed by reflection from the surface. Irregularities in the surface show up as features in the reflected beam. The vertical resolution of optical technique is no better than that of any optical system, namely half the wavelength of light, and the horizontal resolution is comparable. This is a bit of an inconvenience since the wavelength of light, 50 μm, is larger than would be desirable for studying many surfaces.

2.3.3   Electron microscopy

In order to obtain good resolution in both directions, we can resort to electron microscopy. Much important information has been gained in this way. Table 2.3 lists the classification of the various methods available for topographical examination according to the range of vertical heights and spatial wavelengths. Table 2.4 lists the surface characterization techniques.

2.4   Common CMP surface defects

Common CMP defects are dishing, erosion, recess, field loss, and others such as scratches and pits.

Dishing, as its name implies, is a localized defect in which each connecting side is overpolished. It is shown in Figure 2.9a. Erosion is also a localized defect. Instead of one connecting site, erosion is an overpolished site that covers a few more sites (plugs), as shown in Figure 2.9b. Both dishing and erosion are physical defects that can be seen through the microscope. Recess and field loss are overpolished sites, but they are more of an electronic property loss.

Scratches and pits are mainly due to mechanical abrasion and chemical attack, leading to localized physical damage. Examples are shown in Figures 2.9c and 2.9d.

Table 2.4 Summary of Various Methods of Surface Analysis[16]

Acronym	Name	Probe	Detection Signal	Information	Space Resolution	Applications to Tribology
AED	Auger electron spectroscopy	Electron	Auger electron	Elemental analysis of the surface	≈ 50 nm, depth 1.5 mm	Depth direction profile of each element at friction surface (+ ion gun)
CL	Cathode luminescence	Electron	Photon	Illumination spectrum, composition	≈ 5 μm	Lattice defects and compression analysis (+ PMA)
EELS	Electron energy loss, spectroscopy	Electron	Electron (scattered)	Electronic status of very small portion	≈ 10 nm, depth direction 50 μm	Elemental analysis, electron state analysis (+TEM)
ELL	Ellipsometry	Light (laser)	Light (polarized)	Optical constant, film thickness	Several 10 μm, depth direction 50 nm	Stress due to slip, film thickness changes, etc.
EPMA	Electron probe microanalysis	Electron	Characteristic x-ray	EDX, WDX, quantitative analysis	≈ 0.5 μm, depth direction 0.3 to several μm	Analysis of slip products
ESR	Electron spin resonance	Magnetic field	Electromagnetic wave	Unpaired electron (radicals, etc.)	10^{10} spins	Lubricant deterioration and dangling bond analysis
XAFS	X-ray absorption fine structure	X-ray	X-ray absorption	EXAFS (distance to surrounding atoms, number); XANES (atomic value configuration)	0.2 nm	Composition of adsorbed substances

		Neutral	Cation	Atomic configuration		
FIM	Field ion microscopy	Neutral molecule + electric field	Cation	Atomic configuration	—	Three-dimensional structure at atomic level resolution (+ electric field evaporation)
NMR	Nuclear magnetic resonance	Magnetic field	Electromagnetic wave	Molecular bonding condition	> 10 μm	Adsorption condition of lubricant, molecular mobility, surface functional radicals
IR	Infrared absorption spectroscopy	Light (infrared)	Light (infrared)	Vibration condition	Thin film acceptable	Analysis of adsorption performance of gases and lubricants (FT-IP, RAS, ATR, polarization)
LEED	Low energy electron diffraction	Electron	Electron (diffraction)	Molecular bonding condition	Several atom layers below the surface	Gas adsorption at surface (+AES, RHEED)
RBS	Rutherford backscattering spectroscopy	Ion beam	Scattered ion	Vibration condition	> 100 μm	Analysis of slip products, thin film density
OPM	Optical profilometry	Light (laser)	Light (interference)	Surface atomic structure	Subμm, depth direction 0.5, several nm	Imaging to slip surface
RS	Raman scattering	Light (laser)	Light (Raman scattering)	Trace elements, atomic configuration	1 μm	Electronic properties of thin films, etc.
SEM	Scanning electron microscopy	Electron	Electron (secondary scattering)	Surface topography	≈ 1 nm, depth direction 0.3, several μm	Damage forms (*in situ* as well, + EPMA)

(continued)

Table 2.4 Summary of Various Methods of Surface Analysis[16] (Continued)

Acronym	Name	Probe	Detection Signal	Information	Space Resolution	Applications to Tribology
SIMS	Secondary ion mass spectroscopy	Ion beam	Secondary ion	Combination condition	Several μm, depth direction 0.5, several nm	Extremely small quantity element detection including hydrogen
SPA	Surface potential analysis	—	Electric field	Topography	Several 10s of μm	Slip charge
SPM	Scanning probe microscopy	Electric field	Tunnel current	Surface and subsurface formation	$\approx$ 0.1 nm, depth direction $\approx$ 0.1 nm	Various applications including STM and AFM
STEM	Scanning transmission microscopy	Electron	Electron (transmission)	Polarization	$\approx$ 1 nm	Crystallization in very fine regions (+ EDX, + EELS)
TA	Thermal analysis	Thermal energy	Heat dissipation	Microscopic surface structure	—	Adsorption energy, lubricant deterioration, etc.
TEM	Transmission electron microscopy	Electron	Electron (transmission, diffraction)	Shape, structure elements	10 nm, 0.1 nm	Transition cell structure, reaction products, etc.
TOF-SIMS	Time of flight SIMS	Ion beam	Secondary ion	Heat of adsorption, heat of transition	100 μm, depth direction 1 nm	Lubrication distribution, deterioration
TDS	Thermal desorption spectroscopy	Thermal energy	Desorbed atoms, molecules	Imaging	—	Identification of adsorbed and stored materials

TXRF	Total-reflection x-ray fluorescence spectroscopy	X-ray	Fluorescent x-ray	Composition distribution	10^9 atoms/cm,^{2} depth direction several nm	Surface contaminants
UPS	Ultraviolet photoelectron spectroscopy	Light (ultra-violet)	Photoelectron	Adsorption, material decomposition	1 μm	Surface-oriented tribological analysis bonding state
XRD	X-ray diffraction	X-ray	X-ray diffraction	Trace element analysis	Several 100s of μm, depth direction μm	Thin film structure, internal pressure defects, etc. (including neutron diffraction)
XPS	X-ray photoelectron spectroscopy	X-ray	Photoelectron	Chemical composition	100 μm, depth direction several nm	Surface-oriented tribological analysis, bonding state, lubricant film thickness (+ ion gun)

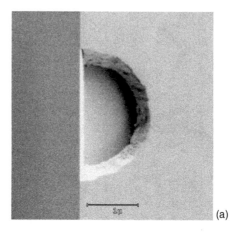

(a)

Figure 2.9a Common CMP defects: dishing.

(b)

Figure 2.9b Erosion. (Continued)

Appendix: ASME standards for surface roughness measurement

ASME Y14.36M-1996; Surface Texture Symbols.

ASME B46.1-1995; Surface Texture (Surface Roughness, Waviness, and Lay).

ISO 1302:1994; Technical Drawings — Method of indicating surface texture.

ISO 4288:1996; Geometrical Product Specifications (GPS) — Surface texture: Profile method — Rules and procedures for the assessment of surface texture.

ISO 12085:1996; Geometrical Product Specifications (GPS) — Surface texture: Profile method — Motif parameters.

ISO 3274:1996; Geometrical Product Specifications (GPS) — Surface texture: Profile method — Nominal characteristics of contact stylus instruments.

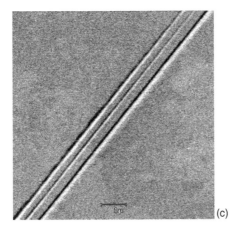

Figure 2.9c Scratches. (Continued)

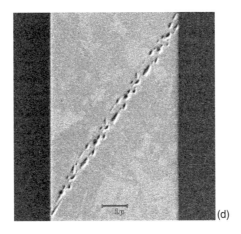

Figure 2.9d Pits. (Continued)

ISO 11562:1996; Geometrical Product Specifications (GPS) — Surface texture: Profile method — Metrological characteristics of phase correct filters.
ISO 13565-1:1996; Geometrical Product Specifications (GPS) — Surface texture: Profile method; Surfaces having stratified functional properties — Part 1: Filtering and general measurement conditions.
ISO 13565-2:1996; Geometrical Product Specifications (GPS) — Surface texture: Profile method; Surfaces having stratified functional properties — Part 2: Height characterization using the linear material ratio curve.
ISO 4287:1997; Geometrical Product Specifications (GPS) — Surface texture: Profile method — Terms, definitions and surface texture parameters.
ISO 5436:1985; Calibration specimens — Stylus instruments — Types, calibration and use of specimens.

ISO 1302:1994; Technical Drawings — Method of indicating surface texture. ISO/TR 14638:1995; Geometrical Product Specifications (GPS) — Master plan: *an ISO Technical Report (TR), not an ISO Standard.*

References

1. J.B.P. Williamson, The shape of surfaces, in *Handbook of Lubrication Theory and Practice of Tribology,* Vol. III, E.R. Booser, Ed., CRC Press, Boca Raton, FL, 1988, p. 7.
2. K.C. Ludema, Friction, in *Handbook of Lubrication Theory and Practive of Tribology,* Vol. 2, E.R. Booser, Ed., CRC Press, Boca Raton, FL, 1988, pp. 31–48.
3. B.J. Hamrock, Fundamentals of Lubrication, Presentation at Union Carbide Corporation, Tarrytown, NY, July 17, 1990.
4. L.D. Wedeven, What is EHD?, *Lubr. Eng.,* 31, 6, 291–296, 1975.
5. C. Barus, Note on the dependence of viscosity on pressure and temperature, *Proc. Am. Acad. U.S.A.,* 19, 13, 1893.
6. W.R. Jones, R.L. Johnson, W.O. Winers, and D.M. Sanborn, Pressure viscosity measurements for several lubricants to 5.5×10^8 newton per square meter (8 $\times 10^4$ PSI) and 149°C (300°F) *ASLE Trans.,* 18, 4, 249–262, 1974.
7. C.H.A. Roelands, Correlation Aspects of the Viscosity–Temperature–Pressure Relationship of Lubricating Oils, University Microfilm, Ann Arbor, MI, 1966.
8. M. Makowska, M. Gradowski, and J. Molenda, Interfacial interactions in tribological contact, *Tribologia,* 3, 254–264, 1998.
9. J. Crawford and P. Psaila, Miscellaneous Additives, in *Chemistry and Technology of Lubricants,* R.M. Mortier and S. T. Orszulik, Eds., VCH, New York, 1992, pp. 160–173.
10. Z.S. Hu, S.M. Hsu, and P.S. Wang, Tribochemical and thermochemical reactions of stearic acid on copper surfaces studied by infrared microspectroscopy, *Tribology Trans.,* 35, 1, 189–193, 1992.
11. J.L. Lauer, B.L. Vleek, and B.L. Sargent, Wear reduction by pyrolytic carbon on tribosurfaces, *Wear,* 162–164, part 1, 498–507, 1993.
12. R. Kempinski, E. Kedzierska, K. Kardaze, L. Wilkanowicz, and M. Konopku, Tribopolymerization-type additives for lubricants, Part I: C_{12}–C_{18} alkyl methacrylates, *Tribologia,* 26(3), 277–298, 1995.
13. V.J. Novotny, X. Pan, and C.S. Bhatia, Tribochemistry at lubricated interfaces, *J. Vac. Sci. Technol.,* 12(5), 2879–2886, 1994.
14. C. Kajdas, P.M. Lafleche, M.J. Furey, J.W. Hellgeth, and T.C. Ward, A study of tribopolymerization under fretting contact conditions, *Lubr. Sci.,* 6(1), 51–89, 1993.
15. *Source:* Selected data from Appendix Table A.1. Physical Properties of Metallic Elements, in E. Rabinowicz, *Friction and Wear of Materials,* John Wiley, New York, 1964, p. 235.
16. K. Nakajima, Surface science and technology in lubrication engineering, *J. Jpn. Soc. Lubr. Eng.,* 31, 6, 363–368, 1986.

chapter three

Friction

3.1 Basics of friction

As mentioned briefly in Chapter 1, friction was first studied by Leonardo da Vinci (1452–1519). He noticed the importance of friction for machines. He designed simple apparatus to study different types of friction and was able to distinguish sliding from rolling friction. Da Vinci stated that the area in contact had no effect on friction, and that if the load of an object was doubled, its friction would also be doubled. Guillaume Amontons (1663–1705) rediscovered the two basic laws of friction after da Vinci. He found that friction was a result of the work done in lifting one surface over the roughness of the other, or of the deforming or the wearing of the other surface. For several centuries after Amontons' work, scientists believed that friction was due to the roughness of the surfaces. Charles-Augustin Coulomb (1736–1806) added to the second law of friction by stating that the effect due to friction is proportional to the compressive force, although for large bodies friction does not follow this law exactly. In 1950, F. Philip Bowden and David Tabor explained the physical meaning of friction. They stated that the true area of contact is a very small percentage of the apparent contact area. The true contact area is formed by asperities. When the normal force increases, more asperities come into contact and the average area of each asperity contact grows. The frictional force is dependent on the true contact area. Modern friction was studied when the atomic force microscope was developed.[1–10] Friction activates the chemical reactions due to bond stretching, bond breaking, reformation, and diffusion. Tribochemical reactions take place owing to the transformation of ions and atoms. As one of two major aspects of chemical-mechanical polishing, friction plays a significant role during CMP. Although there is still a need to obtain a better understanding, here we will review the conventional wisdom on friction during CMP as a guide for future development.

3.2 Laws of friction

Friction is defined as the resistance encountered when one body moves tangentially over another and they are in contact. Friction often embraces two classes of relative motion: sliding and rolling. In industrial processes, frictional energy is usually dissipated as waste heat. The friction force is represented by F and the friction coefficient by μ. Under many sliding conditions, the μ for a given pair of materials and fixed conditions of lubrication is mostly constant. The three laws of friction are:

1. The friction force is proportional to the normal load.
2. The friction force is independent of the apparent area of contact.
3. The friction force is independent of the sliding velocity.

In the first law, the force (F_F) that must be applied to an object to initiate and maintain relative motion is proportional to the applied load (L). The proportionality constant is the coefficient of friction (μ).

$$F = \mu L \qquad (3.1)$$

Virtually in all dry sliding contacts we observe that the frictional force required to initiate motion is more than that needed to maintain the surfaces in the subsequent relative sliding: thus there are two values reported for the coefficient of friction. The *static* coefficient of friction is used in reference to the initial movement of the object from the rest position. In this case, the F mL. The *kinetic* coefficient of friction is used for two surfaces in relative motion. This feature, together with the inevitable natural elasticity of any mechanical system, can often lead to the troublesome phenomenon of stick–slip motion (the displacement of surface materials with time). Displacement increases linearly with time during periods of sticking; when slipping occurs, the deformed surface materials are released. Representative of dry static and kinetic coefficients of friction for various material pairs are found in tribology and physics references;[11] see Table 3.1.

The factors that affect dry sliding friction include:[12]

1. The true area of contact between the sliding surfaces
2. The bond strength between the two bodies at the contact interface
3. The way that the material in and around the contacting region is sheared and ruptured

When two surfaces are in contact, the real contact area (A_r) is dependent on the total area of asperity contact, as shown in Figure 3.1. The area of friction is assumed to be dependent on the A_r and the shear strength (S_s):[6,7]

$$F = A_r S_s \qquad (3.2)$$

Table 3.1 Coefficients of Dry Static and Kinetic
Sliding Friction

Material Pair	μ (Static)	μ (Kinetic)
Aluminum on mild steel	0.61	0.47
Teflon on teflon	0.04	—
Teflon on steel	0.04	—
Copper on mild steel	0.53	0.53
Nickel on nickel	1.10	0.53
Brass on mild steel	0.51	0.44
Brass on cast iron	—	0.30
Copper on cast iron	1.05	0.29
Aluminum on aluminum	1.05	1.4
Cast iron on cast iron	1.10	0.15
Bronze on cast iron	—	0.22

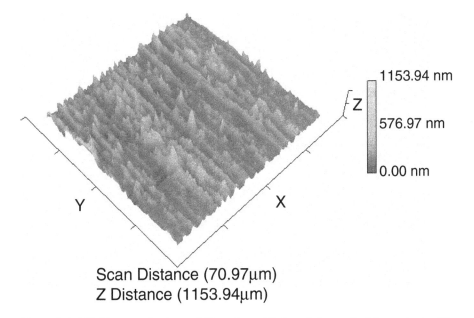

1153.94 nm

576.97 nm

0.00 nm

Scan Distance (70.97µm)
Z Distance (1153.94µm)

Figure 3.1 AFM image of a lapped Si surface. High points are first in contact with another surface. The areas of these high points become the real contact area.

If a load L is carried by an asperity contact area A_r creating the contact pressure P_s, then:[6]

$$\mu = \frac{S_s}{P_s} \qquad (3.3)$$

Although conceptually illustrative, the calculation of the coefficient of friction in practice is much more complex and requires the integration of other interfacial bonding and failure modes which include:[13]

- Area of true contact
- Friction due to adhesion
- Friction due to ploughing
- Friction due to deformation

3.3   Friction due to adhesion

Adhesive fracture is caused by the rupture of interfacial adhesive bonds, which are the summation of all of the component bond strengths within the contact area including van der Waals, ionic, and metallic bonds. The adhesive component of the coefficient of friction (μ_a) has traditionally been estimated from:

$$\mu_a = \frac{F_a}{L} \tag{3.4}$$

where F_a is the interfacial shear strength and L is the load.

3.4   Friction due to ploughing

Ploughing, the formation of a groove in the material, may be caused by either wear debris or asperities within the wear contact. The contribution due to ploughing may dominate the friction force. Currently, there is no reliable model to estimate the contribution to friction force by ploughing.[8] Figure 3.2 shows the typical abrasion, scuffing, and ploughing.[14]

3.5   Friction due to deformation

When we examine a pair of dry sliding surfaces, there is often scoring and surface damage on one or both of them. This makes the direction of relative sliding obvious. The plastic work leading to the ploughing deformation must overcome the initiating or maintaining tangential motion and therefore the frictional force. In a simple picture of surface interactions, the frictional force is simply added to any forces arising from adhesion effects. A schematic of a surface asperity model is shown in Figure 3.3. According to this figure, the frictional force is related to the contact shape, geometry, and depth.

The friction characteristic is the most fundamental factor in tribological processes. In the standard ASTM D4999-89, a method is described of evaluating fluids for their effects on the friction of lubricant-cooled brakes with bronze friction material in combination with steel disks. A chatter and a capacity are generated by the sliding surfaces. Often, surface deformation

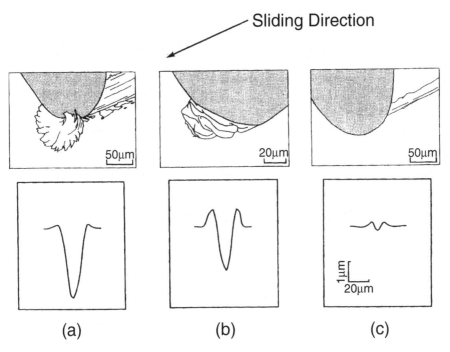

Figure 3.2 Illustration of typical (a) abrasion, (b) scuffing, and (c) ploughing.

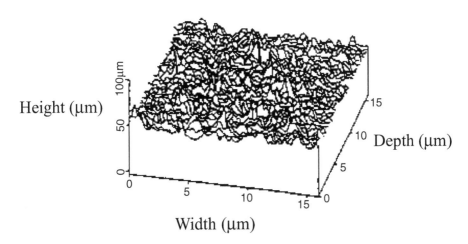

Figure 3.3 A schematic of abrasive asperity.

is found as failure combined with ploughing or adhesion when lubrication was not sufficient. Similarly, a modified tractor from an engine can be used. ASTM D5005-89 describes a tractor with a power-take-off shaft. ASTM D4998-95 describes a similar test method for gear friction and wear for tractor hydraulic fluids. Failure for both is also similar to the previous test method, with which one needs to be cautious.

3.6 Flash temperatures

When two surfaces are sliding against each other, there is an increase in temperature due to an increase in the heat of friction. The local temperature increase at the sliding interface may also be enhanced at the asperity contacts by transient flashes or hot spots. Estimation and measurement of local flash temperatures are both difficult. The simplest way to estimate the highest possible flash temperature is to assume a single asperity carrying the whole load for a brief moment. The interface temperature will be the result of the instantaneous rate of local power dissipation. The temperature rise varies with the sliding speed because the power dissipation depends on velocity. The temperature rise also results from the motion of the heat source over the counterface and must be considered. Under dry sliding conditions in metals, predicted flash-temperature rises may exceed 1000 K. The steady interfacial temperate rises due to the mean frictional power dissipation can add a further hundred degrees or so. This has been confirmed experimentally by thermoelectric and infrared measurements. The contact temperature at the sliding interface is dependent on the applied load at the contact, the time within the contact zone, and the sliding speed.[15] Work by Okabe et al.[16] and Matveevsky[17] has shown that the effectiveness of additive chemistry is also dependent on the geometry of the contact. The contact temperature is therefore a critically important factor in controlling the effectiveness of lubricant additive chemistry, and thus it is important to measure it.[16,18]

Now we estimate the flash temperature. Within a sliding contact, the temperature rises instantaneously. In most conventional cases the temperature of the two frictionally heated surfaces is assumed to be approximately equal and is defined as:[19]

$$T_C = T_S + \mathbf{D}T \tag{3.5}$$

where T_C is the maximum instantaneous surface temperature of the contact, T_S is the quasi-steady-state surface temperature, which is assumed (although incorrectly) to be equivalent to the bulk temperature of the lubricant, and ΔT is the maximum rise of the instantaneous surface temperature above the quasi-steady-state lubricant temperature. The value of ΔT is called the flash temperature.[19] When the value of T_C is sufficient to cause scuffing, this is called the critical contact temperature for scuffing (T_{CR}) and is equal to:

$$T_{CR} = T_S + \Delta T \tag{3.6}$$

As will be discussed in Chapter 5, scuffing is one of the abrasive wear modes that often happens in metals. This equation is known as Blok's critical contact temperature model. The value of T_{CR} is a constant for a given lubricant–metal surface and lubricant–metal surface–atmosphere combination.[19] Interestingly, the value of T_{CR} is independent of the oil viscosity. However, variation of the metal or surface treatment will significantly affect the value of T_{CR}. An example of the interrelationship between steel chemistry and antiwear protection offered by various additives containing sulfur or phosphorous can be found in detail in Huang et al.[20]

Rabinowicz[21] has used the following general rule to estimate the flash temperature of pure dry sliding motion:

$$T_m = \frac{U}{2}\left(\pm \text{ factor of } 3\right) \qquad (3.7)$$

where U is the sliding speed in ft/min and T_m is the flash temperature in °F.

The flash temperature is also related to types of tribomachines. Each type of machine has its specific type of contact geometry and style. For example, on a four-ball lubricant testing machine, the flash temperature is totally different from that of a sliding pin on disk. The ΔT_{max} temperature for a standard four-ball wear contact may be calculated from a Blok equation:[18]

$$\Delta T_{max} = \frac{1.61 n W^{\frac{2}{3}}\mu}{k\left[1+0.627\left(Vr/\alpha\right)\right]^{0.5}} \qquad (3.8)$$

where μ is the coefficient of friction, W is the top ball load (kg), n is the rotational speed in r/sec of the top ball, k is the thermal conductivity (2.604 kg/sec °C), a is the thermal diffusivity of steel (6.045 mm²/sec), and r is the Hertz contact in mm. On this type of machine, the flash temperature (ΔT) has been estimated directly from wear data using the equation:[22,23]

$$\Delta T = \frac{W}{d^{1.4}} \qquad (3.9)$$

where W is the applied load in newtons (N) and d is the mean wear scar diameter (mm). To estimate the flash temperature, we need properties of materials such as hardness, thermal conductivity, specific heat, density, and thermal diffusivity.

3.7 Friction in CMP and post-CMP cleaning

Friction in CMP involves several components: wafers, pads, and slurries (including nanoparticles and other additives). Friction measurements during

and after CMP have been reported for systems from both laboratory and
industrial operations.

Contact during and after CMP is between a wafer and a polymer as a pad
or brush; thus the nature of the contact is predominantly elastic. As we saw
in Section 2.1, the ratio of the elastic modulus E and the hardness H determines
the extent of the plasticity in the contact region as well as the surface topography. For metals, E/H is typically 100 or greater, whereas for many of the
softer polymers (low Es), E/H is only about 10. Thus the contact between
metals and polymers is almost completely elastic except against very rough
surfaces. Another factor that affects the friction of polymers is the strong time
dependence of their mechanical properties: most polymers are viscoelastic and
also show a marked increase of flow stress with strain rate.

In CMP and post-CMP cleaning, one additional significant factor is the
change in the material properties of polymers in water. The water molecules
react with the urethane polishing pad molecules, as is illustrated in Figure
3.4. Water molecules break hydrogen bonds that cross-link the urethane
molecular structure. The breakage weakens the urethane and thus softens
the pad material. This phenomenon starts once a pad is soaked in water and
it continues until it reachs an equilibrium state. Soaking of the polishing pad
is a necessary step. Detailed information about the condition process will be
discussed in Chapter 7.

The coefficients of friction of the polymers used in CMP, as they slide
against wafer materials, commonly lie in the range from 0.2 to 0.6, although
more data are emerging from the CMP community. It is important to know
that when polymeric materials are considered for frictional contact, Amontons' laws are not broadly applicable. The friction coefficient μ varies so
much with applied load, sliding speed, temperature, and slurries that a list
of coefficients of friction for specific CMP processes would be of little value.
Owing to the critical and strict requirements of consistency, friction coefficients can be used as a monitoring factor for CMP processes, as shown in
Figure 3.5. We see that the conditioning can increase the friction coefficient
to a stable level. The friction coefficient also responds to the polishing conditions such as applied load, speed, and nature of the slurry, as shown in
Figure 3.6. Other parameters such as the addition of nanoabrasive particles

Water Molecules Disrupting
the Hydrogen Bond

Figure 3.4 Interactions between water molecules and those of urethane pads.

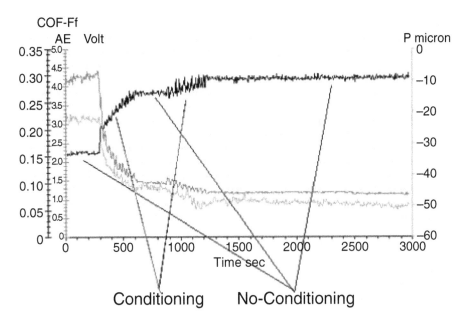

Figure 3.5 Friction as a monitoring factor during CMP. (Courtesy of Jun Xiao.)

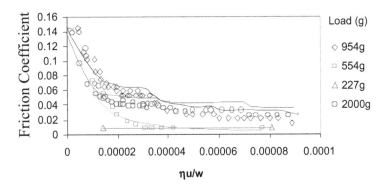

Figure 3.6 During CMP, the friction coefficient is a function of test parameters.

affect friction as well. Owing to friction, the wafer and pad surfaces also change physically and chemically. The physical change can be seen in pad wear. The chemical change is discussed in the tribochemical and chemical wear sections. These can be found in Chapter 5.

The friction of polymers or CMP processes can be attributed to two major sources in mixed modes: deformation involving the dissipation of energy in quite a large volume around the local area of contact, and adhesion originating from the interface between the wafer and the pads (brush). The details of the deformation and adhesion will be discussed in Chapter 5, along with a clear schematic illustration.

References

1. E. Rabinowicz, Polishing, *Sci. Am.*, 218, 91–99, 1968.
2. F.P. Bowden and D. Tabor, *The Friction and Lubrication of Solids*, Clarendon Press, Oxford, 1958, pp. 33–72.
3. T. Ohmi, Trends for future silicon technology, *Jpn. J. Appl. Phys.*, Part 1, 33, 12B, 6747, 1994.
4. F. Preston, The theory and design of plate glass polishing machines, *J. Soc. Glass Tech.*, 11, 214–256, 1927.
5. T.E. Fischer and W.M. Mullins, Chemical aspects of ceramic tribology, *J. Phs. Chem.*, 96, 5690–5695, 1992.
6. H. Tomizawa and T.E. Fischer, Friction and wear of silicon nitride and silicon carbide in water: Hydrodynamic lubrication of low-sliding speed obtained by tribochemical wear, *ASLE Trans.*, 30, 41–46, 1986.
7. T.E. Fischer and H. Tomizawa, Interaction of tribochemistry and microfracture in the friction and wear of silicon nitride, *Wear*, 105, 21, 1985.
8. T.E. Fischer, H. Liang, and W.M. Mullins, Tribochemical lubricious oxides on silicon nitride, in *New Directions in Tribology*, L. Pope, L. Fehrenbacher, and W. Winer, Eds., *Materials Research Society Symposium Proceedings*, 140, 339–344, 1989.
9. T. Fischer, Tribochemistry, *Ann. Rev. Mater. Sci.*, 18, 303–308, 1988.
10. G. Heinicke, *Tribochemistry*, Carl Hanser, München, 1984.
11. K.C. Ludema, Friction, in *Handbook of Lubrication Theory and Practice of Tribology: Vol. 2*, E.R. Booser, Ed., CRC Press, Boca Raton, FL, 1988, pp. 31–48.
12. D. Tabor, Friction — the present state of our understanding, *Trans. ASME J. Lubr. Technol.*, 103, 169–179, 1981.
13. T.A. Stolarski, Basic principles in tribology, in *Tribology in Machine Design*, Industrial Press, Oxfoxd, 1990, pp. 13–63.
14. K. Kato, Wear mechanisms, in *New Directions in Tribology*, I. Hutchings, Ed., World Tribology Congress, London, 1997, pp. 39–56.
15. W.J. Bartz, Tribology, lubricants and lubrication engineering — a review, *Wear*, 49, 1–18, 1978.
16. H. Okabe, M. Masuko, and H. Oshino, Effects of viscosity and contact geometry on tribochemical surface reaction, *ASLE Trans.*, 25, 1, 39–43, 1981.
17. R.M. Matveevsky, Chemical modification of friction surfaces in boundary lubrication, *ASLE Trans.*, 25, 4, 483–488, 1981.
18. A. Sethuramiah, H. Okabe, and T. Sakurai, Critical temperatures in EP lubrication, *Wear*, 26, 187–206, 1973.
19. P.M. Ku, H.E. Staph, and H.J. Carper, On the critical contact temperature of lubricated sliding-rolling disks, *ASLE Trans.*, 21, 2, 161–180, 1978.
20. Y.W. Huang, G.X. Cheng, and J.X. Dong, Studies on the interrelationships between the character of metals and antiwear additives, *Lubr. Sci.*, 2, 3, 253–266, 1990.
21. E. Rabinowicz, *Friction and Wear of Materials*, John Wiley, New York, 1964, p. 35.

22. T. Singh and V.K. Verma, EP activity evaluation of tris(N-arylthiosemicarba-zido)-molybdenum (III) on steel balls in a four-ball test, *Wear*, 146, 313–323, 1991.

23. A. Bhattacharya, T. Singh, V.K. Verma, and N. Prasod, 1,3,4-Thiadazoles as potential EP additivies — a tribological evaluation using a four-ball test, *Tribology Int.*, 28, 3, 189–194, 1995.

chapter four

Lubrication

Lubricants are traditionally used to reduce friction and wear and to provide smooth running and a satisfactory life for machine elements. Lubricants' physical states include liquid, gaseous, and solid states. Different regimes of a lubricating system were recognized starting in the middle of the nineteenth century. The understanding of hydrodynamic lubrication began with classical experiments and celebrated analysis in the 1880s.[1-3] The understanding of boundary lubrication is attributed to Hardy and Doubleday,[4,5] who found that extremely thin films adhering to surfaces were often sufficient to assist relative sliding. In later years, a better understanding and definition of other lubrication regimes between these two extremes were obtained, such as elastohydrodynamic (EHD) and mixed lubrications.

Lubrication in chemical-mechanical planarization (CMP) is still a new concept. It has been reported that EHD and boundary lubrications are possible lubrication regimes.[6] Evidence has shown that boundary lubrication is likely the dominant regime. In this chapter, we will discuss the basics of lubrication concepts that may be useful for CMP applications.

4.1 Lubrication regimes

Lubrication mechanisms have been classically represented by interrelating the coefficient of friction (μ) and operational parameters including lubricant viscosity (η), rotational speed (U), and contact loading (w) using the Stribeck curve shown in Figure 4.1.[7] This curve was first observed by Stribeck in 1902 in fluid lubrication of journal bearings without external pumping. The curve has a minimum, which immediately suggests that more than one lubrication mechanism is involved. The regimes of lubrication are sometimes identified by a lubricant film parameter equal to h/σ, i.e., the mean film thickness over a composite standard deviation of surface heights of the two surfaces. This curve was later described in different lubrication regimes by McKee and McKee in 1929.[8] Figure 4.2 shows the relationship between the friction in the bearing and the expression μ/w, which is called the Hersey number, or Sommerfeld's group. This relationship has been used by engineers to design

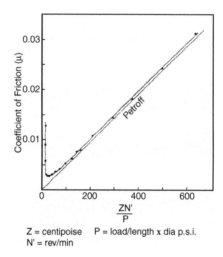

Z = centipoise   *P* = load/length x dia p.s.i.
N' = rev/min

Figure 4.1 A classic Stribeck curve.

bearings and to conduct failure analysis. At least three lubrication regimes are indicated in Figure 4.2: hydrodynamic, elastohydrodynamic (EHD or EHL) including mixed EHD (or EHL), and boundary lubrication.

Hydrodynamic lubrication is characterized by relatively large film thickness, typically greater than 0.25 μm, which is substantially greater than surface asperity heights. In a hydrodynamic regime, surfaces are generally conformal with a positive fluid pressure.

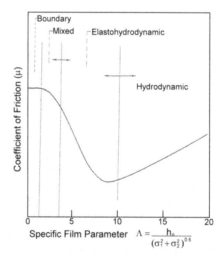

Figure 4.2 The relationship between the friction in the bearing and the expression μ/w, which is called the Hersey number, or Sommerfeld's group.

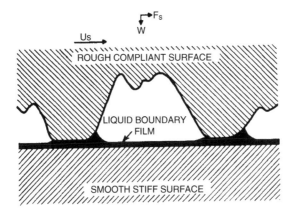

Figure 4.3 Dimensionless film thickness versus speed for different fluids at room temperature.

EHL lubrication is characterized by thin film lubrication with film thickness of approximately 0.025 to 2.5 μm. Although these are thin films, they are greater than the asperity contact heights. EHL is a form of hydrodynamic lubrication in which elastic deformation of the lubricated surfaces becomes significant. EHD lubrication is normally associated with nonconformal surfaces.

Boundary lubrication is characterized by film thicknesses of ≤ 0.025 μm, which is less than the height of the asperity contacts. Mixed film, or EHD boundary lubrication, occurs at the transition from boundary to EHD lubrication. For comparison, the relative sizes of various components of a wear contact are provided in Figure 4.3.[9] Because the solids are not separated by a lubricant, fluid film effects are negligible and there is considerable asperity contact. The contact lubrication mechanism is dominated by the physical and chemical properties of thin surface films of molecular proportions. The properties of the bulk lubricant are of minor importance, and the friction coefficient is essentially independent of fluid viscosity. The frictional characteristics are determined by the properties of the solids and the lubricant film at the common interfaces.

The lubrication behavior of a CMP process has been studied in the past few years.[10–13] The mechanisms are still not quite understood at this stage. In these studies, the effects of viscosity on CMP were investigated, and the average friction coefficient, as a function of relative surface speed and applied load, is estimated for polishing performance. An example of measured friction as a group of parameters is shown in Figure 4.4. This figure indicates that the friction coefficient was the lowest when copper was polished in DI water. The addition of hydrogen peroxide and 0.3 μm abrasive particles increased the friction coefficient steadily. Another example shows the visible change of friction when different slurries are added. This result shows that the increase of the friction coefficient is mainly due to the addition of abrasive particles. We can see that the friction changes during polishing

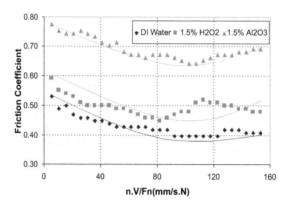

Figure 4.4 Simulation of a Sribeck during CMP.

when the surfaces change. When acidic and alkaline alumina slurries are used, the friction changes as well, as shown in Figure 4.5. The addition of acidic or alkaline agent decreases the friction coefficient, and the lubricating behavior of the copper–CMP system consequently changed. In Figure 4.5, the lowest friction coefficient in acidic and alkaline slurries shifts toward low speed comparing with that in water. The previous study has shown that the CuO film was formed on a copper surface when polished in alkaline slurry.[14,15] The passivation of the copper surface modifies the friction behavior that applies in a range of boundary and mixed lubrication regimes. According to the Pourbaix diagram of a copper–H_2O system, copper would dissolve into Cu^+ when it is polished in an acidic slurry.

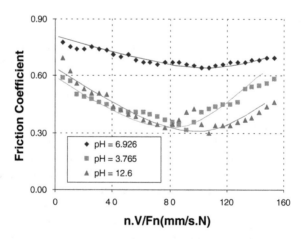

Figure 4.5 Effects of slurry chemistry on friction of CMP.

4.2   Lubrication fundamentals

4.2.1   Hertzian contact

Lubrication involves contact of surfaces. If one body is pressed against another with sufficient pressure, an elastic deformation will result, as shown in Figure 4.6.[16] The pressure generated in this contact is the *Hertzian* stress (σ_H):

$$\sigma_H = \left[\frac{1.5\,L\,E^2}{\pi^3\,r^2\,(1-v^2)^2} \right]^{1/3} \tag{4.1}$$

where L is the contact loading, v is Poisson's ratio (for most materials it is 0.3), r is the contact radius, and E is the modulus of elasticity. Mechanical and material properties in this equation can be found in Chapter 2. Table 4.1 gives the values of typical materials.[17]

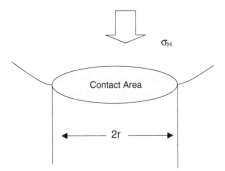

Figure 4.6 Lubrication and contact pressure and elastic deformation.

Table 4.1 Elastic Modulus of Selected Materials

Materials	Elastic Modulus (Gpa)
Aluminum	70.6
Copper	12.98
Tungsten	310
Alumina	300–400
Silicon	100
Silicon oxide	75
Silicon nitride	280–310
Polyethylene — high density	0.5–1.2
Polyurethane (porous)	0.5
PTFE (25% glass fiber)	1.7

Hertz's equation for plastic deformation permits the calculation of the true area of contact (A_r) in the Hertzian contact region:[18]

$$A_r = 2.9[Lr]^{2/3} \tag{4.2}$$

assuming that Poisson's ratio (v) for both surfaces is 0.3.

4.2.2 *Lubricated Hertzian contact*

It has been shown that a ball will elastically deform in a Hertzian contact region that is much smaller than its radius of curvature.[19] The load that produces the deformation is called the Hertzian pressure: it has a parabalic distribution, as shown in Figure 4.7, which is high in the center and low at the edges of contact. Typical Hertzian pressures found in bearing and gear contacts are very high, approximately $1.4 \times 10^9 \, \text{N/m}^2$ (200,000 psi). Note that this is an elastic pressure due to the elastic deformation of the surfaces.

The Hertzian condition of contact is a dominating feature of EHD lubrication. It establishes the overall shape of the contacting surfaces. The overall pathway of a fluid particle would be to enter first an inlet region, then pass through the converging surfaces, and finally exit through a diverging region. The hydrodynamic pressure in this region must be sufficient to separate the surfaces being forced together by the enormous pressure in the Hertzian region, which is of the order of $1.4 \times 10^9 \, \text{N/m}^2$, and the usual hydrodynamic pressures generated in journal bearings are of the order of $7 \times 10^6 \text{N/m}^2$ (1000 psi) or less. The objectives of the following paragraphs are to discuss the inlet region condition and then the process of successful fluid film formation within the contact region for EHD lubrication. This will help the CMP community to understand and optimize the process.

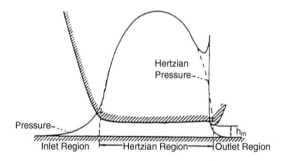

Figure 4.7 Hertzian contact.

4.2.3   Reynolds equation

The practical importance of the mechanism of EHD lubrications lies in the thickness of the fluid film between the surfaces. Its thickness is controlled by the operating conditions of the various operating parameters such as surface velocity, load, and fluid viscosity.

The influence of various operating parameters on film thickness (h_m) is shown by the following Reynolds equation for a line contact formed by a cylinder on a plane:[20]

$$h_m = \frac{2.65\,(\eta_0 U)^{0.7}\,\alpha^{0.54}\,R^{0.43}}{E^{0.03}\,L^{0.13}} \tag{4.3}$$

where h_m = film thickness at the rear constriction of the Hertzian contact
$\quad\quad\quad\eta_0$ = viscosity at atmospheric pressure
$\quad\quad\quad\alpha$ = pressure–viscosity coefficient
$\quad\quad\quad U$ = velocity defined as $1/2(U_1 + U_2)$ where U_1 and U_2 are the
$\quad\quad\quad\quad$ individual velocities of the moving surfaces
$\quad\quad\quad R$ = radius of equivalent cylinder
$\quad\quad\quad L$ = load per unit width
$\quad\quad\quad E$ = elastic modulus of equivalent cylinder (flat surface assumed
$\quad\quad\quad\quad$ completely rigid)

An important feature of EHD lubrication is that the influence of the load on the film thickness is very small. An increase in load merely increases the maximum Hertzian pressure and makes the Hertzian region larger. It does very little to the inlet region where hydrodynamic pressure is generated.

If the film parameter Λ is equal to 1, boundary lubrication and asperity contact occurs; mixed lubrication occurs when $\Lambda = 1 - 3$, and EHD lubrication occurs when $\Lambda = 3 - 10$. The minimum film thickness (h) is dependent on α (the pressure–viscosity coefficient), η_0 (kinematic viscosity), v (entraining velocity), R (effective radius), E (effective elastic modulus), and L = load of a unit line contact. These are interrelated by the following dimensionless parameters:

Dimensionless sliding speed parameter

$$\bar{U} = \frac{\eta_0\,v'}{E'\,R'} \tag{4.4}$$

Dimensionless load parameter

$$\overline{W} = \frac{w}{E' R'} \qquad (4.5)$$

Dimensionless materials parameter

$$\overline{G} = \alpha E' \qquad (4.6)$$

Dimensionless viscosity parameter

$$g_e = (\overline{WU})^{-1/2} \qquad (4.7)$$

Dimensionless viscosity parameter

$$g_v = (\overline{GW})^{3/2} \overline{U}^{-1/2} \qquad (4.8)$$

$$R' = \frac{R R_1}{R - R_1} \qquad (4.9)$$

where R = radius of pump stator
         R_1 = radius of vane profile.

4.2.4   *Non-Newtonian EHD and micro-EHD lubrication*

Fluid viscosity within the inlet region of the Hertzian contact is dependent not only on temperature and pressure, but also on the non-Newtonian rheological properties of the fluid.[21] Hirst and Moore[22] demonstrated that the critical factor is not shear rate, but shear stress within this region.

The critical stress is dependent on the pressure and the molecular size. When the critical stress is exceeded in EHD lubrication, traction coefficients, heat generation, and film thickness will be affected.

The impact of non-Newtonian viscosity behavior on effective film thickness has been successfully modeled with an equation developed by Bair and Khonsari[21] that incorporates the second Newtonian obtained from the Carreau viscosity equation.

Chang and Zhao[23] have compared Newtonian and non-Newtonian analysis of micro-EHD contacts. Based on numerical analysis, it was concluded that the shear thinning effects of the lubricant will produce significant errors if the non-Newtonian behavior is not properly accounted for.[23] Effective pressure–viscosity coefficients have been reported to be dependent on the non-Newtonian behavior of lubricants within the inlet region of the Hertzian contact.[24]

4.2.5 Mixed EHD film lubrication

Mixed EHD film lubrication is encountered during the transition from full EHD film to boundary lubrication. In this region, surface roughness (texture) effects are particularly critical. A common method of accounting for surface roughness effects is through the use of the film parameter (Λ):

$$\Lambda = \frac{h_{min}}{\left(R_{q,a}^2 + R_{q,b}^2\right)^{1/2}} = \frac{h_{min}}{R_{ms}} \tag{4.10}$$

where R_{ms} is the RMS composite asperity amplitude $[R_{q,a}^2 + R_{q,b}^2]^{0.5}$ for surfaces a and b. Figure 4.8 shows the effect of the magnitude of the film parameter on the lubrication mechanism.[2]

Two-dimensional asperity orientation shows a significant effect on film thickness. This is illustrated in Figure 4.9 showing the effect of the surface roughness parameter (γ) on film thickness. The value of γ is defined as the ratio of the correlation length in the direction of flow to the correlation length normal to the flow.[25] Figure 4.9 illustrates that two-dimensional effects are not significant except at low film parameter values (<2). This means that one-dimensional surface roughness analysis greatly overestimates surface roughness effects.

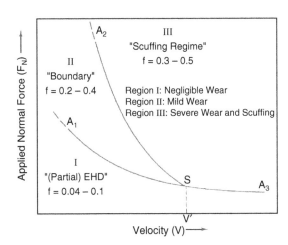

Figure 4.8 The effect of the magnitude of the film parameter on the lubrication mechanism.

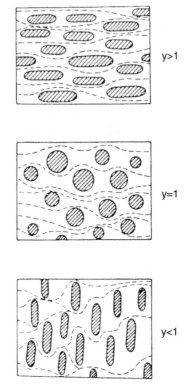

Figure 4.9 Two-dimensional asperity orientation with a significant effect on lubrication film thickness.

The effect of both oil viscosity and surface roughness has been studied.[26] The influences of surface roughness were much greater than the effect of fluid viscosity. The loading capacity before scoring, galling, scuffing, or seizure was highly dependent on surface smoothness. The presence of a single scratch may cause total failure by seizure.

4.2.6 *Petroff's law*

Hydrodynamic lubrication is characterized by conforming surfaces so that the load is carried over a large area. Petroff, in his classic analyses of lubrication, assumed that the lubricant sticks to the surfaces of the journal bearings, that the shaft is concentric, and that the velocity gradient is equal to U/c, where U is the surface speed of the shaft and c is the film thickness. The frictional torque (T) is:

$$T = \frac{\eta R U}{c} A \qquad (4.11)$$

where η is viscosity, *R* is the radius of the bearing, and *A* is the "wetted" area.[27a]

The coefficient of friction (μ) is:

$$\mu = 2\pi^2 \frac{\eta n}{P} \cdot \frac{1000}{\delta}$$ (4.12)

where *n* is the rotational speed in 1000 revolutions per second, *P* is the load per unit area, and δ is the radial clearance ratio 1000 *c/R* in thousandths of an inch per inch of shaft diameter.

This can be rewritten in the more familiar form of the Stribeck curve as shown in Figure 4.2.

$$\mu \; \alpha \; \frac{Z'n}{Ph_o}$$ (4.13)

where *Z'* is *Zähigkeit*, the German term for viscosity, *n* is the surface speed, *P* is the applied pressure, and h_o is the average film thickness. The low values of ZN/Ph_o differ from the line shown in Figure 4.1, and are often called the Petroff line. Petroff did not account for the temperature dependence of viscosity, especially at higher temperatures, or the effects of surface asperity variation at low clearances.[8,27b]

4.3   Boundary lubrication in CMP

Previous studies have indicated that no hydrodynamic lubrication occurs during CMP.[28–31a] There is always a physical contact between the wafer and the polishing pad asperities. In the following section, we will see that there is enough evidence to prove interactions between a wafer and a pad. The boundary lubrication associated with tribochemical interactions plays a dominant role. In order to understand the mechanisms of boundary lubrication in CMP, the physical, electrochemical, and mechanical processes of interfaces must be considered. The mechanisms can be classified into the following categories based on the surface physical chemistry of materials involved during CMP.

4.3.1   Passivation of metals

4.3.1.1   Tungsten (W) CMP
Metals commonly used in CMP include tungsten, aluminum, copper, and tantalum. To date there are several mechanisms reported. The passivation of tungsten was first proposed by Kaufman in 1991[31b] and has been widely used in guiding slurry design in the CMP community.[7] Figure 4.10 illustrates the passivation and removal procedures. This model involves the formation

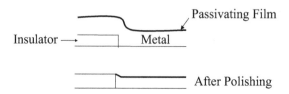

Figure 4.10 Passivation and removal procedures during tungsten (W) CMP.

of a blanket passivating layer on the surface of the tungsten due to the oxidizing nature of the slurry. According to this model, mechanical abrasion from the slurry and the polishing pad subsequently removes the passivating layer. This bare metal, when exposed to the oxidizer, immediately repassivates. The abrasion–repassivation process is hypothesized to continue until a layer that has a much lower polish rate than tungsten is formed. This mechanism requires that all the tungsten removed be in an oxidized state. The constituents of the aqueous polish slurry used (by the same group of researchers) were potassium ferricyanide (an oxidant), ethylenediamine (a complexant for tungsten), and potassium dihydrogen phosphate (a buffering agent, slurry pH6).[7] The proposed etching reaction was:

$$W + 6Fe(CN)_6^{-3} + 3H_2O \rightarrow WO_4^{2-} + 6Fe(CN)_6^{-4} + 8H^+ \qquad (4.14)$$

The competing passivation reaction was proposed as:

$$W + 6Fe(CN)_6^{-3} + 3H_2O \rightarrow WO_3 + 6Fe(CN)_6^{-4} + 6H^+ \qquad (4.15)$$

It was suggested that the relative rate of each process was dependent on the pH and the rotation rate due to the hydrogen ion formation and transport rates.

A parallel study of the electrochemical behavior of tungsten during CMP was conducted to elucidate the role of passivating layer formation and oxidation.[32] The polishing rate was measured with and without potentiostatic control of different slurries with changing oxidizers and complex regents. Results showed that the tungsten oxidation rates during polishing are one to two orders of magnitude below the polishing rate. The polishing rates, based on measured currents during anodic potentiostatic control, are much higher than expected. The polish rate is unaffected by cathodic potentiostatic control. These results indicate that the removal mechanisms of tungsten during CMP might not require a blanket passivating film or the oxidation of all removed tungsten. This observation indicates that both passivation and direct chemical removal might coexist.

4.3.1.2 Copper (Cu) CMP

In the case of copper (Cu) CMP, engineers have found that the process is less stable; often copper has more defects than other metals have. In order to understand the Cu CMP mechanisms, it is necessary to know the surface chemistry occurring during polishing. Figure 4.11 shows the results of energy dispersive x-ray (EDX) chemical analysis of slurries before and after polishing.[23] The amount of certain elements is shown as the intensity of the peaks. In Figures 4.11a and 4.11c (left), the amounts of certain elements present are shown as the intensity of the peaks before and after polishing. Figures 4.11b and 4.11d (right) show aggregated fumed silica particles. The spectra of Figure 4.11a show the presence of aluminum (from the aluminum grit of the carbon film). The concentration of copper is an artifact from apertures in the microscope. Figure 4.11 establishes a baseline for the analysis. The intensity of the Cu peak significantly increases after CMP is done, as shown in Figure 4.11c. In Figure 4.11, the existence of potassium (K) is visible between 3 and 4 keV. The potassium is added to disperse the nano-abrasive particles. The chemical mapping analysis results of slurries after polishing are shown in Figure 4.11b and 4.11d. After polishing, the silicon mapping is shown in Figure 4.11d (located at the upper right corner), where the increased darkness represents a higher concentration of silicon. Inside the particle, the concentration of silicon is apparently high. The potassium and copper were mapped as well, and the results are shown in Figure 4.11d (lower left corner) and in Figure 4.11d (lower right). Two elements are homogeneously distributed both inside and outside of the silica particle. Potassium was a common addition as KOH to silica slurry; there is no concentrated K on silica particles. Copper is found here distributed evenly outside and inside the silica particles. In all of these photos, there is no metallic copper found. This indicates the atomic level removal or dissolution of copper taking place during CMP.

The passivation of copper is almost immediate when it is exposed to oxygen.[9,33] As is apparent in Figure 4.11, when copper was polished with high pH fumed silica, no metallic particles remained, nor was any large copper-oxides–wear debris detected in the polished slurry.[27b] Using x-ray analysis, the ratios of the changing concentration of copper and oxide were calculated. Results show that the dissolved form of copper is most likely Cu_2O. During the friction measurement, it was observed that the color of the copper during polishing changed with slurries. The color of the copper surface is dark when it is polished with water, and this color is also particularly apparent when polished with H_2O_2. However, the copper has a bright, shining, and reddish color when it is polished with alumina slurries. We also know that the CuO is naturally dark in color, while the Cu_2O appears more similar to copper's original color. Therefore we believe that Cu_2O is most likely to form during polishing with alumina slurry and CuO most likely forms during polishing with water or H_2O_2. The x-ray results indicate that Cu_2O might form during the use of a high-pH fumed silica slurry. This is consistent with our assumption for Cu polishing with alumina slurry. According to Table 4.2, the crystal structure for Cu_2O is octahedral.

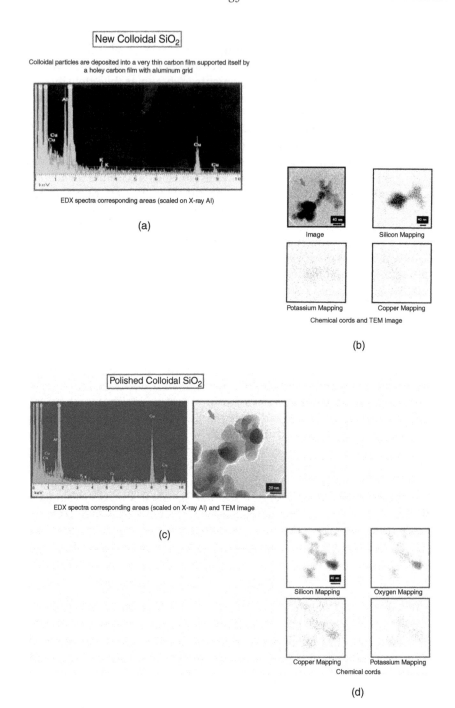

Figure 4.11 The energy dispersive x-ray (EDX) chemical analysis of slurries before and after polishing.

Table 4.2 Material Properties of Copper and Its Oxides

	Cu	Cu$_2$O	CuO
Specific gravity[34]	8.95	6.0	6.4
Hardness (HV)[35]	<130	~150	—
Tensile strength (Mpa)[10]	220–455	—	—
E (MPsi)[10]	18.8	—	—
Dissolution rate (nm/min)[36,37,38]	8	—	—
Polishing rate (nm/min)[10,39]	10–40	—	—
Crystal structure[40]	Cubic	Octahedral cubic	Monoclinic
Color[18]	Reddish metal	Reddish metal	Black

It is clear that if a harder product such as Cu$_2$O is formed, scratches on Cu will appear.

4.3.2   Adhesion of oxides

Oxide CMP, namely the chemical-mechanical polishing of silicon dioxide, is a process for planarization of ILD (interlayer dielectric) and STI (shallow trench isolation). Materials currently used in such manufacturing are TEOS, thermal oxide, BPSG, low K, and so on. Oxide CMP was adapted from glass polishing, where the effect of water on glass was reported as a softening agent.[41–44] The general reactions for the interaction of glass (Si-O) and water (H$_2$O) are:[39]

$$\text{-Si-O-Si}^- + \text{H}_2\text{O} = \text{-Si-OH} \tag{4.16}$$

or

$$(\text{SiO}_2)\text{x} + 2\text{H}_2\text{O} = (\text{SiO}_2)_{\text{x-1}} + \text{Si(OH)}_4 \tag{4.17}$$

The pH value of a slurry has pronounced effects on the removal rate of Si, as shown in Figure 4.12. Figure 4.13 is the solubility of amorphous silica as a function of pH.[45] Based on these results for the removal rate of Si CMP and the dissolution rate of amorphous Si as a function of pH, the removal rate is a function of pH, as shown in Figure 4.14. The ongoing chemical reactions occurring on the silicon surface during CMP using colloidal suspension silica have been reported by Pietsch et al.,[46,47] as is illustrated in Figure 4.15. Using IR, they found that Si(111) and Si(100) surfaces were dominated by hydrogen, which made the surfaces strongly hydrophobic and chemically stable in air because of the passivation occurring. The hydrogen peak was at pH = 11. At a very high concentration of OH$^-$ and at pH >11, the formation of larger, more stoichiometric suboxide clusters is favored because of the OH$^-$ concentration-dependent increase of the electrochemical potential for the direct oxidation of Si^{0+} → Si^{4+} (the Nernst equation). The quartzlike surface layer is likely formed and removed,[48] and it is consistent

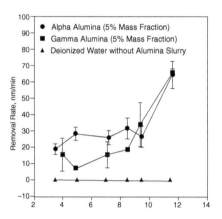

Figure 4.12 The effect of pH value of a slurry on removal rate of Si during CMP.

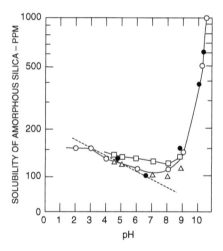

Figure 4.13 The solubility of amorphous silica as a function of pH.[45]

with our chemical analysis results, in which the oxygen concentration is increased. Thus surface chemistry plays an important role in material removal.

The chemical interactions are associated with the slurry chemistry and process conditions. This can be analyzed using the x-ray energy dispersion chemical analysis that is summarized in Table 4.3. Four samples were compared. The pH effects on Si as well as the mechanical cutting of Si and glass are shown. The atomic concentrations of elements detected are listed. A significant amount of oxygen remains on the silicon surface after it is polished in a pH = 11.5 slurry. Compared with the concentration of oxygen in the glass, as shown in Table 4.3, there is relatively less oxide on the polished silicon surface. This disparity in oxide presence is evidence of the oxidation

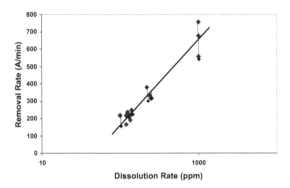

Figure 4.14 The removal rate is a function of pH.

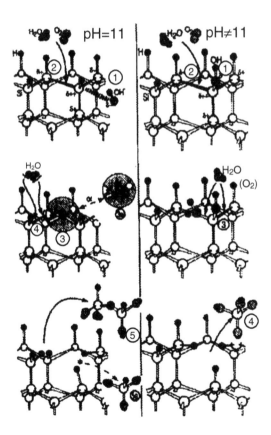

Figure 4.15 The ongoing chemical reactions occurring on the silicon surface during CMP using a colloidal suspension.[46,47]

Table 4.3 X-Ray Chemical Analysis

Atom Percent	Si Polished in pH = 3.5	Si Polished in pH = 11.5	Dry Cut Si	Glass Surface
Si-k	66.56	20.17	98.84	18.05
O-k	3.42	59.31	0.09	78.15
C-k	29.87	19.53	1.07	—
Al-k	—	—	—	0.87
Na-k	0.15	1.00	0.00	7.98

of silicon induced through polishing. The carbon manifested itself on the polished Si surface, but not on the glass surface. This indicates the transfer of carbon from the polishing pad during polishing. That the carbon is not visible on the glass surface may have two causes. One is that C has a very low signal. The second is the high bonding strength of Si=O. The energy provided during pad–wafer contact is not enough to break the bonds to form new carbon-containing bonds. The fact of carbon transfer from a polishing pad to the metal layer surface of a wafer during CMP has been shown in our previous studies.[11,20]

During CMP, silicon undergoes oxidation to form SiO_2.[49,50] A series of relatively simple extrapolations and estimations have led to conclusions that CMP of oxide materials, as it is currently practiced by industry, is a synergistic process of the formation of an adhesive layer on an oxide wafer surface and around abrasive particles and the removal of these layers by pulling of particles from the wafer surface, in a snowball effect.[42] This means that adhesion between the constantly regenerating hydrated wafer surface and the surface of hydrated slurry particles leads to surface material removal. A schematic diagram of this is Figure 4.16.[51]

4.3.3   Tribochemical reactions

More detailed study indicates that transfer wear actually exists during CMP.[52] The interactions between a copper wafer and urethane polishing pads were characterized to investigate the effects of friction on removal mechanisms of a polishing system of copper interconnected wafers in water. *In situ*

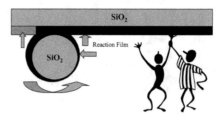

Figure 4.16 The schematic diagram of oxide snow ball removal mechanism.

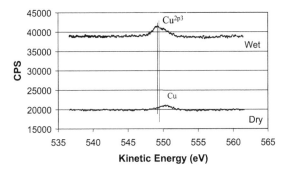

Figure 4.17 XPS analysis of *in situ* characterization of polished copper.

characterization of polished copper and urethane were conducted using the Auger and XPS analysis techniques, as shown in Figure 4.17. In this figure, the Cu and its oxide peaks were found on the polishing pad and the C was found on the Cu wafer. These results indicated that, owing to the stimulation of friction, the molecules from the pad transferred to the copper surface, and the oxidized copper surface was transferred to the urethane surface. Without friction, however, such transformation did not occur, and passivation of the copper surface took place. This evidence proves a tribochemical CMP mechanism. In addition to the formation and removal of a passivation layer, a transformation layer is formed during CMP owing to friction stimulation. This layer is found on both copper and pad surfaces with different chemical bonds.

4.4   Lubrication behavior in post-CMP cleaning

Like a CMP process, the post-CMP cleaning is a tribological system as well.[53-55] Figure 4.18 shows the friction coefficient as a function of the average relative surface speed. This figure has two curves: one is for a dry (moist) brush and one is for a brush soaked with water. The wet brush apparently has a lubricating behavior as the friction decreases with speed. Owing to a lack of fluid, the dry brush on the contrary has an increasing friction against speed.[56] In Figure 4.19, the friction coefficient decreases as the load increases. Similar experiments are reported in References 13, 14, and 57. The cleaning brush is highly elastic. In order to remove particles that remain on the wafer surfaces, a boundary lubrication and EHD lubrication regimes are recommended.

4.5   Additive classification for slurry design

Lubrication design has a long history in the tribology community in oil-based lubricants for reducing friction in engines, motors, and hydraulic systems. Scientists and engineers in the CMP community have designed slurries using different chemical additives. Many patents are available. There is, however,

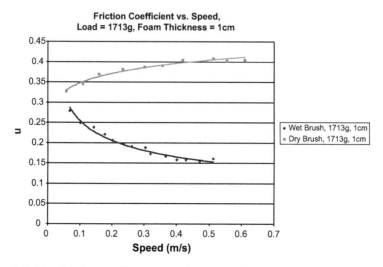

Figure 4.18 The friction coefficient as a function of average relative surface speed during post-CMP cleaning.

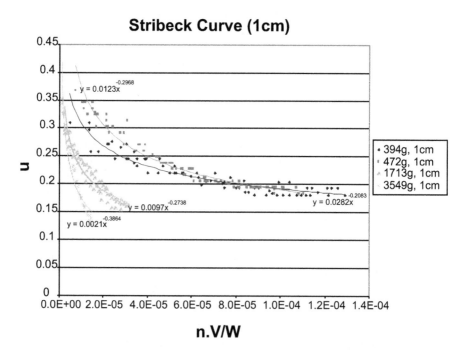

Figure 4.19 The friction coefficient as a function of applied load during post-CMP cleaning.

Table 4.4 Common Additive Types and Functions

Additive	Purpose	Function
Friction modifier (FM) (Refs. 59 and 60)	Reduce friction under near-boundary conditions	Adherence of polar materials to metal surfaces
Antiwear additive	Reduce wear	Formation of film on metallic contacting surfaces
High pressure (HP) additive	Prevent indents, galling, scoring, and seizure	Formation of low-shear films on metal surfaces at the wear contact

no detail reported. In Table 4.4 we list the parameters considered for lubrication additives. We hope this can be a guide for CMP practitioners. For details, please refer to Reference 58.

References

1. N.P. Petrov, *Friction in Machines and the Effect of the Lubricant*, Inzh. Zh., St. Petersburg, 1883, Vol. 1, pp. 71–140; Vol. 2, pp. 227–279; Vol. 3, pp. 377–436; Vol. 4, pp. 535–564.
2. B. Tower, Second report on friction experimets (experiment on the oil pressure in a bearing), *Proc. Inst. Mech. Eng.*, 58–70, 1885.
3. O. Reynolds, On the theory of lubrication and its application to Mr. Beauchamp Tower's experiments, including an experimental determination of viscosity of olive oil, *Philos. Trans. R. Soc. London*, Ser. A, 177, 157–234, 1886.
4. W.B. Hardy and I. Doubleday, Boundary lubrication — the paraffin series, *Proc. R. Soc. London*, Ser. A, 100, 25–39, March 1, 1922.
5. W.B. Hardy and I. Doubleday, Boundary lubrication — the temperature coefficient, *Proc. R. Soc. London*, Ser. A, 101, 487–492, Sept. 1, 1922.
6. J. Tichy, J. Levert, L. Shan, and S. Danyluk, Contact mechanics and lubrication hydrodynamics of chemical-mechanical polishing, *J. Electrochem. Soc.*, 146, 4, 1523–1528, 1998.
7. R. Stribeck, *Zeit. Ver. Deut.*, 46, 180, 1902.
8. S.A. McKee and T.R. McKee, Friction of journal bearings as influenced by clearance and length, *Trans. Am. Soc. Mech. Engrs.*, 51, 161–171, 1929.
9. R.S. Fein, Boundary lubrication, in *Handbook of Lubrication: Theory and Practice of Tribology, Vol. II, Theory and Design*, CRC Press, Boca Raton, FL, 1988, pp. 49–68.
10. G. Grover, H. Liang, and C.K. Huang, Effect of slurry viscosity modification on oxide and tungsten CMP, *Wear*, 212, 10–13, 1998.
11. H. Liang, E. Estragnat, J. Lee, K. Bahten, and D. McMullen, Proc. Sixth Int. Conf. CMP for ULSI Multilevel Interconnection (CMP–MIC), Santa Clara, CA, 2001, p. 266. First presented at the Lake Placid CMP Conference, August 2000.
12. G.H. Xu and H. Liang, Tribological Behavior of Copper Chemical-Mechanical Polishing, Proc. Sixth Int. Conf. Solid-State and IC Tech., Shanghai, Oct. 22–25, 2001, pp. 851–854.
13. H. Liang and G.H. Xu, Anti-Lubrication Behavior of Cu CMP, Proc. Eighth Int. VMIC, Sept. 25–26, 2001, Santa Clara, CA, 2001, pp. 137–140.

14. H. Liang, J.M. Martin, and R. Lee, IEEE and TMS, *J. Elec. Matls.*, April 2001.
15. H. Liang and G. Xu, Late news article, Proc. Sixth Int. Conf. CMP for ULSI Multilevel Interconnection (CMP–MIC), Santa Clara, CA, 2001, p. 186.
16. U.L. Möller and U. Boor, *Lubricants in Operation*, VDI Verlag, Düsseldorf, Germany, 1996, pp. 1–28.
17. Goodfellow Catalog, http://www.goodfellow.com.
18. E. Rabinowicz, *Friction and Wear of Materials*, John Wiley, New York, 1964, p. 35.
19. H.R. Hertz, *J. Reine Angew. Math.*, 92, 156–171, 1881.
20. D. Dowson, Elastohydrodynamics, *Proc. Inst. Mech. Engrs.*, 182, Part 3A, 151–167, 1967–1968.
21. S. Bair and M. Khonsari, An inlet zone analysis incorporating the second Newtonian, *J. Tribology*, 118(2), 341–343, 1996.
22. W. Hirst and A.J. Moore, Non-Newtonian behavior in elastohydrodynamic lubrication, *Proc. R. Soc. Lond.*, Ser. A, 337, 101–121, 1974.
23. L. Chang and W. Zhao, Fundamental differences between Newtonian and non-Newtonian micro-EHL results, *J. Tribology*, 117, January 29–35, 1995.
24. M. Aderin, G.J. Johnson, H.A. Spikes, and G. Capuriccio, The elastohydrodynamic properties of some advanced non-hydrocarbon-based lubricants, *Lubrication Engineering*, 633–638, August 1992.
25. N. Patir and H.S. Cheng, Effect of surface roughness orientation on the central film thickness in EHD contacts, *Proc. Fifth Leeds Lyon Symp. Tribol.*, 1978, pp. 15–21.
26. S. Anderson and E. Salas-Russo, The influence of surface roughness and oil viscosity on the transition in mixed lubricated sliding steel contacts, *Wear*, 174, 71–79, 1994.
27a. A. Cameron, *Principles of Lubrication*, Longmans Green, London, 1966.
27b. M. Pourbaix, *Atlas of Electrochemical Equilibria in Aqueous Solutions*, NACE, Houston, TX, 1975.
28. G.H. Xu and H. Liang, Tribological Behavior of Copper Chemical-Mechanical Polishing, Proc. Sixth Int. Conf. Solid-State IC Tech., Oct. 22–25, 2001, Shanghai, pp. 851–854.
29. H. Liang, and G.H. Xu, Lubrication behavior in CMP, Proc. Seventh ICSICT, Oct. 18–21, 2004, Beijing, P.R. China, IEEE Press.
30. H. Liang and G.H. Xu, Anti-Lubrication Behavior of Cu CMP, Proc. Eighth Int. VMIC, Sept. 25–26, 2001, Santa Clara, CA, 2001, pp. 137–140.
31a. H. Liang, and G. Xu, Lubricating behavior in chemical-mechanical polishing of copper, *Scripta Materialia*, 46, 5, 343–347, 2002.
31b. F.B. Kaufman, D.B. Thompson, R.E. Broadie, M.A. Jaso, W.L. Guthrie, D.J. Pearson, and M.B. Small, Chemical-mechanical polishing for fabricating patterned W metal features as chip interconnects, *J. Electrochem. Soc.*, 138, 11, 3460–3465, 1991.
32. D. Stein, D. Hetherington, T. Guilinger, and J.L. Cecchi, *In situ* electrochemical investigation of tungsten electrochemical behavior during chemical-mechanical polishing, *J. Electrochem. Soc.*, 145, 9, 3190–3196, 1998.
33. V. Brusic, M.A. Frisch, B.N. Eldridge, F.P. Novak, F.B. Kaufman, B.M. Rush, and G.S. Frankel, Copper corrosion with and without inhibitors, *J. Electrochem. Soc.*, 138, 8, 2253, 1991.
34. G.S. Braely and H.R. Clauser, *Materials Handbook*, 3rd ed., McGraw-Hill, Columbus, OH, 1991.

35. C.T. Lynch, *CRC Handbook of Materials Science*, Vol. I, *General Properties*, CRC Press, Boca Raton, FL, 1991.

36. M. Hoshino, H. Suehiro, K. Kasai, and J. Komeno, *Jpn. J. Appl. Phys.*, 32, L392, 1993.

37. V. Brusic, D. Scherber, F. Kaufman, R. Kistler, and C. Streinz, Electrochemical approach to Au and Cu CMP process development, Proc. First Symp. on CMP in IC Device Manufacturing, Electrochemical Society, San Antonio, Texas, October 6–11, 1996.

38. C. Streinz, T. Myers, and C. Yu, The Effect of Slurry Chemistry on Both Blanket and Patterned CMP Performance, Proc. Fifth Int. Symp. CMP in IC Device Manufacturing, Pennington, NJ, 1996, p. 159.

39. H. Hirabayashi, M. Higuchi, M. Kinoshita, H. Kaneko, N. Hayasaka, K. Mase, and J. Oshima, Proc. First Int. CMP for VLSI/ULSI Multilevel Interconnection Conference (CMP-MIC), Santa Clara, February 1996, p. 22.

40. *CRC Handbook of Chemistry and Physics*, 68th ed., 1987–1988, p. B-89.

41. H. Landis, P. Burke, W. Cote, W. Hill, C. Hoffman, C. Kaanta, C. Koburger, W. Lange, and S. Luce, Integration of chemical-mechanical polishing into CMOS integrated circuit manufacturing, *Thin Solid Films*, 220, 1–2, 1–7, 20, November, 1992; Switzerland, Conference Information 19th International Conference on Metallurgical Coatings and Thin Films. San Diego, CA, April 1992, pp. 6–10,

42. L. Holland, *The Properties of Glass Surface*, Chapman & Hall, London, 1964.

43. T. Izumitani, in *Treatise on Materials Science and Technology*, M. Tomozawa and R. Doremus, Eds., Academic Press, New York, 1979, p. 115.

44. M. Tomozawa, K. Yang, H. Li, and S.P. Murarka, Basic Science in Silica Glass Polishing, Mat. Res. Soc. Symp. Proc., S.P. Murarka, A. Katz, K.N. Tu, and K. Maex, Eds., 337, 89, 1994; Advanced Metallization for Devices and Circuits — Science, Technology and Manufacturability Symposium. Mater. Res. Soc. 1994, pp. 89–98, Pittsburgh, PA, Conference Information: Advanced Metallization for Devices and Circuits — Science, Technology and Manufacturability, San Francisco, CA, 4–8 April 1994.

45. R. Iler, *The Chemistry of Silica, Solubility, Polymerization, Colloid and Surface Properties, and Biochemistry*, John Wiley, New York, 1979, p. 42.

46. G.J. Pietsch, G.S. Higashi, and Y.J. Chabal, Chemo-mechanical polishing of silicon: Surface termination and mechanism of removal, *Appl. Phys. Lett.*, 64, 3115, 1994.

47. G.J. Pietsch, Y.J. Chabal, and G.S. Higashi, The atomic-scale removal mechanism during chemical-mechanical polishing of Si(100) and Si(111), *Surf. Sci.*, 331–333, 395–401, 1995.

48. P.O. Hahn, M. Grundner, A. Schnegg, and H. Jacob, Correlation of surface morphology and chemical state of Si surfaces to electrical properties, *Appl. Surf. Sci.*, 39, 436–456, October 1989.

49. H.J. Larsen-Basse and H. Liang, Oral presentation, MRS Spring Meeting, 1999, San Francisco, CA, 2002.

50. E. Estragnat, G. Tang, H. Liang, S. Jahanmir, P. Pei, and J.M. Martin, Experimental investigation of mechanisms of Si CMP, *J. Elec. Matls.*, 33, 334–339, 2004.

51. J. Larsen-Basse and H. Liang, private discussions.

52. H. Liang, Th. Le Mogne, and J.M. Martin, Interfacial transfer between copper and polyurethane in CMP, *J. Elec. Matls.*, 31, 8, 2002.

53. H. Liang, E. Estragnat, J. Lee, K. Bahten, and D. McMullen, Mechanisms of Post-CMP Cleaning, Proc. Sixth Int. Conf. CMP for ULSI Multilevel Interconnection (CMP–MIC), Institute of Microelectronics Inter-Connection, March 7–9, 2001, Santa Clara, CA, pp. 266–272.

54. K. Bahten, D. McMullen, H. Liang, E. Estragnat, T. Zhang, and J. Lee, The Mechanism of Particle Removal and Brush Mechanics in Post-CMP Cleaning Applications, Proc. Sixth Int. Conf. CMP for ULSI Multilevel Interconnection (CMP-MIC), Institute of Microelectronics Inter-Connection, March 7–9, 2001, Santa Clara, CA, pp. 266–274.

55. H. Liang, E. Estragnat, J. Lee, K. Bahten, and D. McMullen, Interfacial Forces in Post-CMP Cleaning, MRS2001, Proc. MRS-CMP2001 — Advances and Future Challenges, 671, 2001, M7.4-M7.5.

56. R. Stribeck, Characterizatistics of plain and roller bearings, *Zeit. Ver. Dent. Ing.*, 46, 1341–1348, 1902.

57. Prof. Mori and Prof. Liang, private conversations, July 2000.

58. H. Liang, G. Totten, and G. Webster, Lubrication and tribology fundamentals, in *Manual on Fuels, Lubricants, and Standards: Application and Interpretation*, G. Totten, R. Shah, and S. Wesbrook, Eds., 2003, chap. 35; Marcel Dekker, ASTM Int., W. Conshohocken, PA, pp. 909–962.

59. G.J. Schilling and G.S. Bright, Fuel and lubricant additives, *Lubrication*, 63, 2, 13–24, 1977.

60. S. Hironaka, Working mechanisms of additives in lubricating oils, *Sosei to Kako*, 36, 413, 579–585, 1995–1996.

chapter five

Wear in CMP

Wear is one of the most important parameters in evaluating the CMP process. Wear in CMP is evaluated as the material removal rate (MRR). The primary wear mechanisms that occur in CMP are adhesive wear, abrasive wear, electrochemical wear, tribochemical wear, and fatigue wear on both wafer and pad surfaces. In this chapter, we will first introduce basic wear concepts. We will then discuss wear in polishing and in conditioning. Throughout the text, we show examples of CMP failure due to wear.

5.1 Basics of wear

5.1.1 Adhesive wear

Conventionally, adhesive wear occurs when two interacting surfaces are not sufficiently lubricated, which results in the adhesive transfer or removal of near-surface material. Surface adhesion is dependent on the nature of the contacting surface. Wear material transfer in this region has been called "solid-phase welding," which occurs at asperity contacts.[1] A schematic of adhesive wear is shown in Figure 5.1.[2] Adhesive wear is related to the materials and slurries used. The amount of slurry can shift contact from the boundary lubrication regime to the EHD/micro-EHD regime where surfaces are more separated.

The classic expression used to describe quantitatively the value of the adhesive wear debris removed for a given load L, on a material of hardness H, over the sliding distance X, for a dry contact, is Archard's equation:[3,4]

$$V = \frac{k\,L\,x}{H} \tag{5.1}$$

where k is the wear constant. Typical values of the wear constant (for a dry contact) are provided in Table 5.1.[5]

Although Archard's equation is frequently misused, and the physical meaning of the wear constant k is misunderstood or misused, it is still true

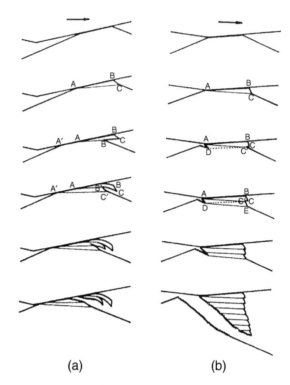

(a)   (b)

Figure 5.1 Adhesive wear; (a) a thin flake-like wear particle and (b) the adhesive transfer of a wedge-like wear particle.

that material hardness is an important variable in the wear process. Hardness, however, is a complex value that is dependent on position and time, temperature, sliding speed, and environment. Localized hardness variation can affect the transition between friction and wear.[5] Meng and Ludema[6] have reviewed the various wear models that have been reported to date and have found that there are no general equations for a wide range of practical wear prediction problems (Table 5.1).

Table 5.1 Wear Constant (k) for Various Sliding Combinations

Material Pair	Wear Constant ($k \times 10^3$)
Copper on copper	32
Copper on low-carbon steel	1.5
Low-carbon steel on copper	0.5

Source: H.C. Meng and K.C. Ludema, *Wear*, 181–183: 443–457, 1995.

5.1.2 *Abrasive wear*

Abrasive wear occurs when one of the contacting surfaces is harder than the other. The observed wear behavior involves plastic deformation and material displacement during ploughing or smearing. This results in surface-initiated fatigue spalling. Figure 5.2 shows the schematic of an abrasive wear model.[7] According to this figure, the abrasive wear volume can be evaluated as:

$$V = (Lx \tan \theta) \div (\pi p) \tag{5.2}$$

In abrasive wear by hard particles we often find either two-body abrasive wear or three-body abrasive wear, as shown in Figure 5.3. Two-body wear is caused by hard protuberances on the counterface, while in three-body wear hard particles are free to roll and slide between two sliding surfaces. Wear rates due to three-body abrasion are generally lower than those due to two-body abrasion. Various mechanisms of material removal in these two cases differ only in relative importance. Slurry erosion belongs to the abrasive wear category. Erosion is caused by hard particles sticking to the surface entrained in a flowing liquid.

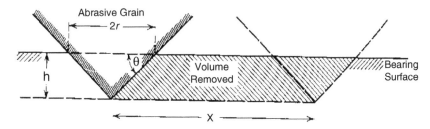

Figure 5.2 An abrasive wear model.

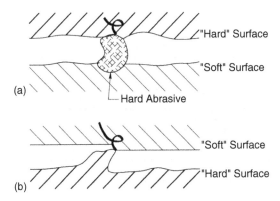

Figure 5.3 Abrasive wear by hard particles often means either two-body abrasive wear or three-body abrasive wear; (a) three body and (b) two body.

5.1.3   *Tribochemical wear*

Unlike chemical reactions, tribochemical reactions are triggered by frictional force, resulting in wear. Tribochemical wear can be seen as one type of corrosive wear. When the corrosion is activated by mechanical interactions between the contacting surfaces, it produces activated surface sites and localized high temperatures sufficient for chemical reaction. Tribochemical wear involves surface charging of electrons, surface passivation, and surface film removal processes.

Tribochemical wear has been observed during CMP using Auger and XPS analysis.[8,9] One instance was found during Cu CMP, while a nonequilibrium oxide CuO was observed after CMP. This is because of frictional stimulation, so that the CuO formed and remained in slurry, as discussed in Chapter 4, Section 4.3.3. The second evidence was found on both the Cu surface and the pad surface after CMP, as discussed in Section 4.3.3. Owing to friction, elements from the pad surface transferred onto the wafer surface and vice versa.

5.1.4   *Fatigue wear*

Fatigue wear by pitting and spalling is caused by induced subsurface shear stresses that exceed the critical shear stress of the material. Fatigue wear has been reported in pad wear during CMP, as shown in Figure 5.4. Fatigue life, or rather pad life, is influenced by the ratio of slurry film to surface roughness. Although fatigue failure on wafers has not been reported, high localized stresses at surface defect sites, asperities, dents, and material inhomogeneity will likely introduce fatigue failure.

5.1.5   *Scuffing*

In his original work published in 1959, Borsoff[10] reported the physical and chemical absorption of surfaces by lubricating additives. Such additives affect the continual intermittent contact between two surfaces.[11] Scuffing takes place when wear takes places during a short-time contact under high

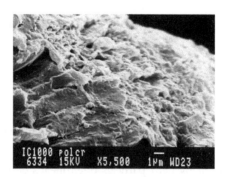

Figure 5.4 An example of pad fatigue wear.

temperature. Since CMP requires continuous material removal, scuffing is considered minor.

5.2   Slurry particles

This topic is yet to be understood. The current understanding is at the micrometer scale. When particles shrink to nanometers, properties and performances change significantly. Here we discuss existing principles of abrasive particles in tribological applications.

In abrasion or erosion, particles with lower hardness cause less wear than harder particles. When particles are much harder than a surface, the exact value of their hardness matters much less. The relative wear rates by two-body abrasion are associated with a wide range of metals and ceramics, abraded by various types of grit particle. The wear rate becomes much more sensitive to the ratio of abrasive hardness H_a to the surface hardness H_s when H_a/H_s is less than 1. Table 5.2 lists the bulk material hardness values for common abrasive particles.[12]

The wear rate depends strongly on the shapes of the particles. Angular particles cause greater wear than rounded ones. The angularity of abrasive particles is difficult to identify and quantify. The simplest description of shape is based on measurement of the perimeter and area of a two-dimensional projection of the particle, usually generated by microscopy. A roundness factor F can then be defined as the ratio between the area A of the projection and the area of a circle with the same perimeter P as the projection:

Table 5.2 Hardness of Selected Materials

Abrasive Materials	Material Hardness (HV)
Diamond	6000–10000
Boron carbide, B_4C	2700–3700
Silicon carbide	2100–2600
Alumina (corundum)	1800–2000
Quartz (silica)	750–1200
Soda–lime glass	500
Fluorite, CaF_2	180–190
Aluminum	25–45
Copper	40–130
Gold	30–70
Silver	25–80
Tungsten	260–1000
Tantalum carbide	1800–2450
Tungsten carbide, WC	2000–2400
Tantalum nitride	1200–2000
Tantalum diboride	2450–2910
Tungsten diboride	2400–2660

$$F = \frac{4\pi A}{p^2}$$

(5.3)

When $F = 1$, the projection is a circle. The more the outline of the particle departs from circular, the smaller the value of F.

The particle size has strong influence on the removal rate during CMP, as shown in Figure 5.5.[13] Without the changing chemical reactions, the smaller the particles, the less wear.

5.3   Pad wear during polishing

It has been reported that wear in polymers depends on their surface roughness. Deformation wear (via the abrasive and fatigue wear mechanisms) occurs when surfaces are rough. Interfacial wear (via adhesive or transfer wear mechanism) dominates when surfaces are smooth.[14] Deformation wear includes abrasive and fatigue mechanisms. Interfacial wear includes adhesive or transfer wear. Transfer wear is defined as pulling and adhesion from the opposite surface. It involves adhered debris and transferring elements from the opposite surface. Deformation wear occurs when surfaces are rough. Interfacial wear dominates when surfaces are smooth. This conclusion applies to polymeric material wear in dry sliding conditions. In CMP applications, however, the polymeric pads are used against liquid slurry and wafer materials. The scale of contact between wafer and pad, and polishing conditions, is more complicated. Figure 5.6 shows morphologies of polishing pads before and after polishing on a laboratory machine through SEM analysis. Figure 5.6a shows pad surface as conditioned, and Figure 5.6b shows the same after polishing. The conditioned pad surface shows pieces of cutting debris and microcutting (abrasion) (Figure 5.6a). In Figure 5.6b, the

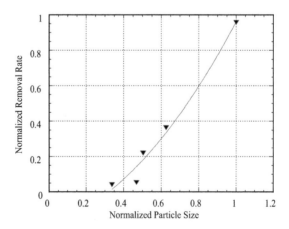

Figure 5.5 Effects of particle size on abrasive wear.

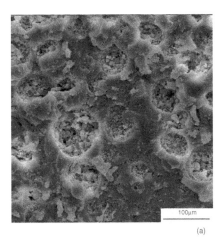

(a)

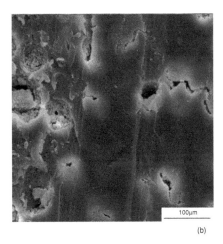

(b)

Figure 5.6 Morphologies of polishing pads before and after polishing on a laboratory machine through SEM analysis. (a) Pad surface as conditioned and (b) the same after polishing.

surface is obtained after polishing for 30 minutes. These features indicate pad material deformation, hardening, and debris adhesion.

Pad wear occurs during pre-CMP conditioning and during CMP. Pad life is also related to the chemical attack of water molecules, transferring wafer elements. Pad wear is mainly dominated by mixed wear modes.

References

1. A. Dorinson and K.C. Ludema, *Mechanics and Chemistry in Lubrication*, Elsevier, New York, 1985.

2. T. Kayaba and K. Kato, Theoretical estimation of abrasive wear resistance based on microscopic wear mechanism, in *Wear of Materials*, K.C. Ludema, W.A. Glaeser, and S.K. Rhee, Eds., ASME, 1979, pp. 45–56.

3. R. Holm, *Electric Contacts*, H. Gerbers, Stockholm, 1946.

4. J.F. Archard, Contact and rubbing of flat surfaces, *J. Appl. Phys.*, 24, 981–988, 1953.

5. D.A. Rigney, The rules of hardness in the sliding behavior of materials, *Wear*, 175, 63–69, 1994.

6. H.C. Meng and K.C. Ludema, Wear models and predictive equations: Their form and content, *Wear*, 181–183, 443–457, 1995.

7. E. Rabinowicz, *Friction and Wear of Materials*, 2nd ed., John Wiley, New York, 1995, p. 192.

8. H. Liang, J.M. Martin, and R. Lee, Influence of oxides on friction during Cu CMP, IEEE and TMS, *J. Elec. Mater.*, 30, 4, 391–395, 2001.

9. H. Liang, J.M. Martin, and T.L. Mogne, Interfacial transfer between copper and polyurethane in CMP, *J. Elec. Mater.*, 31, 8, 872–878, 2002.

10. V.N. Borsoff, On the mechanism of gear lubrication, *ASME Trans., J. Basic Eng.*, 80D, 79, 1959.

11. A. Cameron, The role of surface chemistry in lubrication and scuffing, *ASLE Transactions*, 23, 4, 388–392, 1980.

12. I.M. Hutchings, *Tribology — Friction and Wear of Engineering Materials*, CRC Press, Boca Raton, FL, 1992, p. 137.

13. H. Liang, J. Li, and R. Erck, *Tribological Behavior in Chemical-Mechanical Planarization*, Proc. Third Int. Conf. on Surf. Eng., Chengdu, China, Oct. 10–13, 2002, pp. 95–101.

14. B.J. Briscoe and D. Tabor, The effect of pressue on the fictional properties of polymers, *Wear*, 34, 29–38, 1975.

chapter six

Force transmission

The role of the polishing pad consists of two primary processes: (1) the transmission of force to the wafer surface and (2) the distribution of slurry to the wafer surface. This chapter will examine the role the pad plays in generating a force field across the wafer surface during the planarization process. Although many elements come into play during this process, and different polishing circumstances affect these elements differently, the nature of these parameters and their interaction will be examined in a general fashion. This chapter will review the following topics: the inherent nature of a typical pad polymer and its response to force impulses, the nature of cellular solids and the roles typical pad pores can play, the characteristics of asperities and their role in force transmission, and finally some considerations about abrasives and their likely participation in material removal.

6.1 CMP force transmission considerations

At the heart of the force response behavior of the polishing pad is the nature of the pad itself. First the basic mechanical properties of solid polymers will be examined, after which we will review the role of pore structures; and finally we briefly discuss the manufacturing issues.

6.1.1 Mechanical properties of polymeric materials

The simplest view of the behavior of rubberlike materials is that of an energy spring as it displays elastic or Hookean force response behavior; this is represented by the equation $F = -kx$, where F represents the responding force and x the displacement from equilibrium (Figure 6.1a). The long chain nature of polymeric materials, however, presents further complications over a simple spring model; and their behavior is additionally characterized by the number of possible ways that a given configuration can be expressed. As such, they are more often referred to as probability springs (or entropy springs), as illustrated in Figure 6.1b. The thermodynamics of deformation

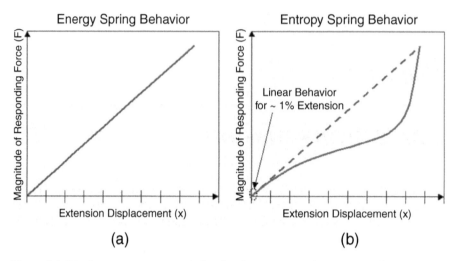

Figure 6.1 Displacement–response behavior for energy and entropy springs.

and the statistical theory underlying the elasticity of a molecular network will not be covered here.

6.1.2   General elastic behavior

To characterize a material's response to a force, certain conventions have developed in the field of mechanics. An elastic material will respond to a force by deforming:

$$F \sim \Delta x \quad \text{(Hooke's law)} \tag{6.1}$$

But the material dimension response depends on its original length, so to get a characteristic that is more a function of the material's intrinsic properties, we look at the ratio of the extension to the original length:

$$F \sim (\Delta x)/x \tag{6.2}$$

The force needed to cause an extention will also depend on the cross-sectional area:

$$F \sim A(\Delta x)/x \tag{6.3}$$

The proportionality constant is called Young's modulus (E), and the response equation for an elastic material is thus:

$$F = EA(\Delta x)/x \tag{6.4}$$

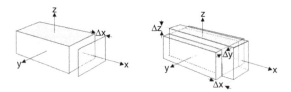

Figure 6.2 Material segment dimensions for a *standard segment* and Poisson's ratio.

We refer to the impressed force per unit area as stress and the responding deformation per unit length as strain. If we divide the equation (6.4) by the area, we obtain:

$$F/A = E(\Delta x)/x \qquad (6.5)$$

Thus stress = E(strain).

A second property of elastic materials (which also comes from Hooke's law) is an attempt to conserve volume. In other words, as the material stretches in one direction, it tends to contract in a perpendicular direction. The amount of contraction depends on both the width and the strain (see Figure 6.2):

$$\Delta y/y \sim (\Delta x)/x \qquad (6.6)$$

The constant of proportionality is called Poisson's ratio and is represented by sigma:

$$\Delta y/y = -\sigma(\Delta x)/x \qquad (6.7)$$

Poisson's ratio is conventionally always positive and is less than 0.5 in magnitude. For a homogeneous isotropic material, these two constants (E and σ) completely specify the elastic behavioral response.[1]

Forces and displacements are directional quantities (vectors), and thus we need to keep track of the magnitudes and directions of our input and responding variables. Since the forces and the responses generally have components in all directions, we describe them through arrays of vectors (tensors).

When the strain response of a material is examined, it is convenient to have notation to keep track of the directions, which is usually done as follows: $\Delta x = e_{xx}x$ (the double subscript on the strain coefficient is to allow a tracking convention for all directions). If the strain response of the material is homogeneous across the material, this coefficient will be a constant. If the material is not homogeneous, however, this coefficient could vary in magnitude, depending its location. We will therefore need to keep track of the changes as we change location, with the stress coefficient rate of change given by $e_{xx} = d(\Delta x)/dx$. This represents compressive (or tensive) type strains. The representation of the shear type strains can be done via directional indices to represent perpendicular directions. In order to keep from encompassing rotational motion into the representation, the cross terms are written $e_{xy} = e_{yx} = \int (d\Delta y / dx + d\Delta x / dy)$.

In general, we have a complete characterization of the strain response if we have the nine strain coefficients:

$$e_{ij} = \begin{Bmatrix} e_{xx} & e_{xy} & e_{xz} \\ e_{yx} & e_{yy} & e_{yz} \\ e_{zx} & e_{zy} & e_{zz} \end{Bmatrix} \tag{6.8}$$

In a homogeneous material, the cross terms are equal ($e_{zx} = e_{xz}$), which means we need only six coefficients to allow full characterization.

For each event we will have a stress tensor S_{ij}, representing the applied stress, and since a stress will produce a strain, we can formulate the relationship as follows:

$$S_{ij} = \Sigma_{k,l} C_{ijkl} e_{kl} \tag{6.9}$$

The coefficent array C_{ijkl} relates the stress tensor to the strain tensor and is itself a tensor (called the *tensor of elasticity*). Fortunately for a homogeneous material the 81 terms in the tensor of elasticity reduce through symmetry considerations to only three,

$$C_{xxxx} = \lambda \tag{6.10}$$

$$C_{xxyy} = 2\mu \tag{6.11}$$

$$C_{xyxy} = 2\mu + \lambda, \tag{6.12}$$

where λ and μ are called the *Lame elastic constants*. This set of equations can be related to the two basic material parameters E and σ as follows:

$$C_{xxxx} = E / (1 + \sigma)\{1 + \sigma / (1-2\sigma)\} \tag{6.13}$$

$$C_{xxyy} = E / (1 + \sigma) \{ \sigma / (1-2\sigma)\} \tag{6.14}$$

$$C_{xyxy} = E / (1 + \sigma) \tag{6.15}$$

Since the sum of the changes in the stress tensors for each directional component change is equal to the force density causing that stress,

$$F_i = \Sigma_j \, dS_{ij} / dx_j, \tag{6.16}$$

we can derive the motion of the material in response to the impressed force. The path to the force response starts by knowing the impressed displacement. From the displacement, the strain can be determined. Once the strain has been found, the stress can be computed (Equation 6.9). Knowing the resultant stress, the force density can be determined (Equation 6.16), which identifies how the displacement will be changing. The closed form equation is far from convenient, but with some simplifying assumptions, an acceptable analytical expression can be derived,

$$\rho \, d^2\mathbf{u}/dt^2 = (\lambda + \mu)\mathbf{\nabla}(\mathbf{\nabla}\cdot\mathbf{u}) + \mu\mathbf{\nabla}^2\mathbf{u}, \qquad (6.17)$$

where $\mathbf{u}$ is used to represent the general displacement vector.

This equation of motion has similarities to wave equations in electromagnetism, and we can follow the analogy to represent our material equation of motion as a composite of two waves. The first wave is a compressional wave moving at a speed of $\sqrt{(\lambda + 2\mu)/\rho}$; the second wave is a transverse (shear) wave that propagates with a velocity of $\sqrt{\mu/\rho}$. The solution of the equations of motion, for materials of shapes common to our use, is very complex. Most approaches have sought to use numerical methods to generate approximations using energy minimization principles.

This analytical solution review is tractable only for very limited assumptions, such as homogeneity and linearly elastic behavior (not to mention excluding variations that are time- or temperature-dependent). The first deviation that must be examined is the elastic linearity assumption for polishing pads. Polymers, in general, show behavior that lies between that of an elastic solid and a viscous fluid. The term viscoelastic has been applied to this behavior.

6.1.3   Viscoelastic behavior

The term *viscoelastic* refers to the response properties of some materials that lie between (or constitute a combination of) the properties of an elastic solid and a viscous liquid. An elastic solid is characterized by its ability to deform its shape under the action of a force while internally storing the energy of the deformation. When the force is removed, the original shape returns with the release of the stored energy. A viscous liquid, however, has no definable shape to begin with and responds to a force by flowing with no retrievable energy in it during the process.[2]

These are the characteristic extremes of low molecular weight materials. As more mass and more structure are added to the materials, their behavior is shifted between the extremes. Both mass and structure play a role because we are interested in the material's response to a deforming force.

The characteristic balance of elasticity and viscosity can be shifted in polymers through either the temperature or the frequency of the deforming force; one works on the effects of structure, the other on the effects of inertia (mass).[2]

One method that attempts to characterize the viscoelastic behavior is the force-deformation measurement. In this approach, a force point is applied

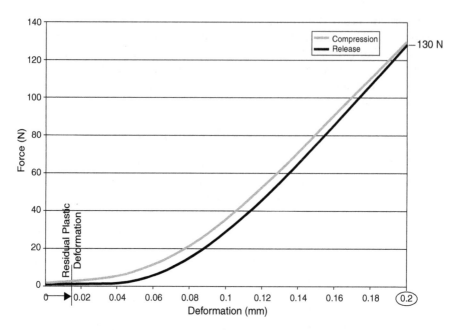

Figure 6.3 Force-deformation curve for characterizing a polishing pad.[3]

to the material, and the subsequent deformation is recorded. The force is applied in steps, and the response is recorded until a preset force or deformation is achieved. Then the process is reversed — the force being withdrawn in steps and the deformation response recorded. An example of such a measurement is shown in Figure 6.3.[3] Typically the compression and the release curves trace out different paths, with a residual plastic deformation at the conclusion, indicative of the viscous aspects of the material. [2]

The behavioral equations for viscous material are rather different from those of elastic materials. The stress that is developed in a viscous fluid is proportional to the velocity gradient (since a liquid has no definable shape, and displacement does not render a storage effect). The coefficient of viscosity η is the proportionality constant relating stress with the velocity gradient:

$$\sigma = \eta \, dV/dz \tag{6.18}$$

The expression for the shear in the xz plane is:

$$\sigma_{xz} = \eta(dV_x/dz + dV_z/dx) \tag{6.19}$$

The velocity components, however, can be written as the time variation of the displacements, giving:

$$\sigma_{xz} = \eta\{d(dx/dt)/dz + d(dz/dt)/dx\} \tag{6.20}$$

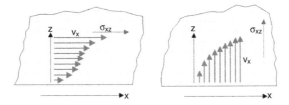

Figure 6.4 Velocity gradients of σ_{xz} and σ_{zx}.

Rearranging the derivatives, we find:

$$\sigma_{xz} = \eta\{d(dx/dz)/dt + d(dz/dx)/dt\} \qquad (6.21)$$

$$\sigma_{xz} = \eta\{d/dt[(dx/dz) + (dz/dx)]\}. \qquad (6.22)$$

But $[(dx/dz) + (dz/dx)]$ is just the shear strain e_{xz}, so we can write $\sigma_{xz} = \eta de_{xz}/dt$, which means that the shear stress is equal to the time rate of change of the shear strain. An illustration of velocity gradients for two directions is represented in Figure 6.4.

In trying to model the response behavior of a viscoelastic material, the next step is to identify what kind of combination of elastic and viscous behavior is realistic. One approach would be to postulate a linear combination of the two. If we represent the shear modulus by G, the two shear terms could be written as

$$(\sigma_{xz})_E = Ge_{xz} \qquad (6.23)$$

$$(\sigma_{xz})_V = \eta de_{xz}/dt \qquad (6.24)$$

The total stress–strain response could then be written $(\sigma_{xz}) = f((\sigma_{xz})_E, (\sigma_{xz})_V)$, such as:

$$\sigma_{xz} = (\sigma_{xz})_E + (\sigma_{xz})_V = Ge_{xz} + \eta de_{xz}/dt \qquad (6.25)$$

The two classic models involve either a parallel or a linear combination of the two terms. The first is called the Kelvin or Voigt model; it involves a spring of modulus E_k and a dashpot with viscosity η_k in parallel, as shown in Figure 6.5a. The immediate response of the spring element is held back by the viscous drag of the dashpot, providing a *capacitive charging* type response. The relaxation response is similarly delayed. In this configuration, the stresses of each element are identical (parallel), and the total strain on the system will be just the sum of the individual element strains:

$$\sigma_{total} = (\sigma_1) + (\sigma_2) = E_{Kel}e + \eta_{Kel}de/dt \qquad (6.26)$$

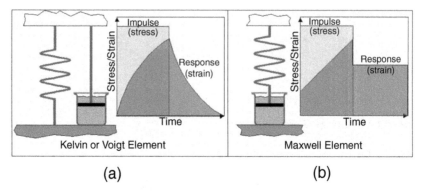

Figure 6.5 Linear viscoelastic classical model elements.

The descriptive formula can be written:

$$\sigma = E_{Kel}e + \eta_{Kel}de/dt \qquad (6.27)$$

If we represent (η_{Kel}/E_{Kel}), which has units of time, by τ', the expression can be shown to deliver:

$$e = (\sigma/E_{Kel})[1 - \exp(-t/\tau')] \qquad (6.28)$$

This exponential behavior can be seen in the curve of Figure 6.5a, which has a peak value of (σ/E_{Kel}) and a charging or discharging decay with the characteristic time of τ'. Although this configuration does a decent job of modeling creep behavior, it cannot represent the typical stress relaxation phenomena. This is because the dashpot only responds to the rate of change of the stress; after a step function in stress (no further change in stress), the second term in Equation 6.27 would be zero, and the strain would be a constant $(\sigma = E_{Kel}e)$ without any time-dependent relaxation behavior.

The second classical model (the Maxwell model) configures the two elements in series, as seen in Figure 6.5b. In the series configuration, the stress is the same for each element, and the total strain is the sum of the individual strains. The descriptive formula can be written:

$$de/dt = (1/E_{Max})d\sigma/dt + \sigma/\eta_{Max} \qquad (6.29)$$

Now the stress relaxation response $(de/dt = 0)$ reduces Equation 6.29 to

$$d\sigma/\sigma = -(E_{Max}/\eta_{Max})dt \qquad (6.30)$$

Representing (η_{Max}/E_{Max}) by τ (referred to as the relaxation time), the equation:

$$\sigma = \sigma_0 \exp(-E_{Max}/\eta_{Max})t \qquad (6.31)$$

can be written as:

$$\sigma = \sigma_0 \exp(-t/\tau) \tag{6.32}$$

The mathematical behavior suffers from several defects, including the description of creep as $d\sigma/t = 0$, which projects viscous flow ($de/dt = \sigma/\eta_{Max}$). In reality, the stress relaxation characteristics do not go to zero over long time periods, and they are often more complicated than a single exponential decay.

Since neither model adequately describes the behavior of real viscoelastic materials, a combination of the classic elements is often made to gain closer representation. The most common configuration is called the standard linear solid[4] configuration, and it is illustrated in Figure 6.6. A more accurate representation of actual behavior can be obtained by a composite of multiple elements of the *standard linear solid* configuration into a multi-element model (Figure 6.7) with an array of coefficients for each element.

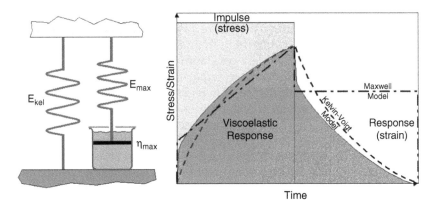

Figure 6.6 Elements of the standard linear solid (SLS) and response curve.

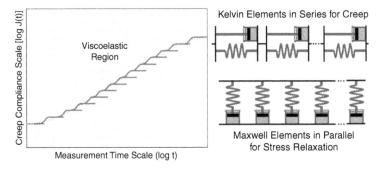

Figure 6.7 Multielement model approximation.[2]

6.1.4   *Creep and stress relaxation characteristics for viscoelastic materials*

The chemical and mechanical structures of polymers can be very complex, but their behaviors can be understood in general terms by observing their responses to stress impulse and relaxation. The strain response of these materials is broadly represented by three components: the immediate elastic deformation, the time-dependent elastic deformation, and the viscous flow response. Since cross-linked polymers, such as are typical of polishing pads, show negligible viscous flow behavior, this third aspect will be generally ignored. A useful approach to observing the creep response is to track what is referred to as *creep compliance*, which is the ratio of the strain $e(t)$ over the stress σ. The approach of characterizing the response by immediate and long-term effects fundamentally introduces the metric of time. Plotting the creep compliance against the time over which the creep response is allowed to take place (most clearly illustrated with logarithmic scales for both axes) provides a characteristic fingerprint of the polymer behavior. Ignoring a flow response that could develop for very long times in non-cross-linked polymers, three regions emerge: a region of low compliance that is relatively constant in time for rapid times (high-frequency measurements); a region of high compliance that is relatively constant in time for slow times (low-frequency measurements); and an intermediate region that bridges the two (see Figure 6.8[3]). The low-compliance stiff region corresponds to glassy solid behavior, the high-compliance flexible region corresponds to rubberlike solid behavior, and viscoelastic behavior is represented by the transitional region. The best characterization for creep is the midpoint of the viscoelastic region, designated by the representational time τ' (retardation time). This transition point is a single number used to represent the behavioral characteristics of the viscoelastic polymer, identifying (as well as one number can) the transition between the two extreme response modes. Although it is often referred

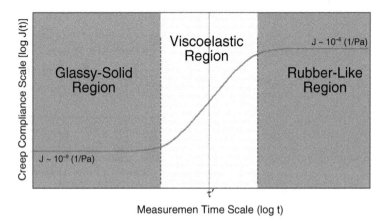

Figure 6.8 Creep compliance curve showing polymer creep–response behavior.[2]

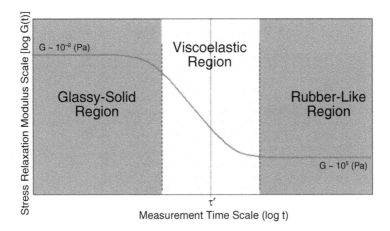

Figure 6.9 Stress relaxation modulus curve illustrating polymer stress–response behavior.[2]

to as *the glass transition point* or *the glass transition temperature*, the knee of the curve would be a better representation of the transition into the glassy solid behavior.

There is an equivalent curve that can be drawn to illustrate the relaxation behavior of a polymer, with three equivalent regions (Figure 6.9[2]). One reason that the midpoint is referred to as *the glass transition temperature* is that a thermal characterization of polymers can be made that exhibits the same type of transition (glassy solid → rubberlike); the transition occurs because of temperature alterations of the material rather than stress impulse time.

Chemical-mechanical planarization presents elements of both time and temperature to the pad material, and therefore the response of the material to both stresses should be understood.

The aspect of stress impulse frequency is an area that has had little exposure in the past, so it is worth some consideration here. The starting point should be a translation of the planarization actions into appropriate time scales to enable a perspective of the typical stress impulse frequencies encountered during polishing.

We now understand that what is nominally considered to be an elastic or a viscous elastic material may well behave like a glassy solid if the impulse frequency is too high for the material to respond elastically. The characteristic stress impulses applied by various components of the planarization process are illustrated in Figure 6.10. The polishing pad used for the process, the CMP tool configuration, the circuit die size, and the different sections of actual device layout in the circuit will contribute a unique impulse frequency during polishing. The stress frequencies encountered during planarization are illustrated for some typical industry values, but the actual values set up in a process should be calculated individually.

Once the creep response of a pad has been determined, it can be superimposed on the characteristic stress impulse graph to identify which

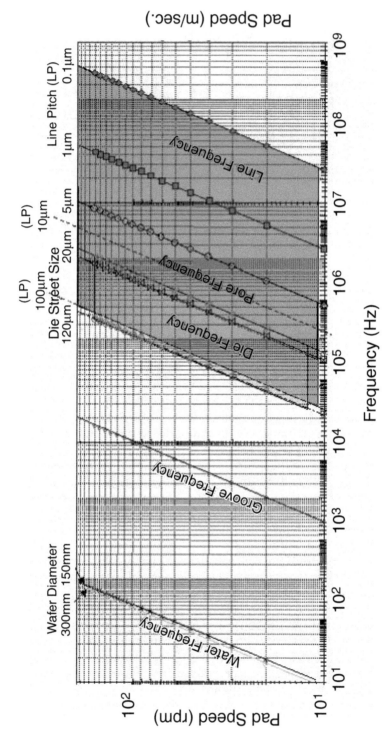

Figure 6.10 Characteristic stress impulses in CMP (rotary and linear).

impulse generators will experience the pad as a glassy solid and which will see a viscoelastic material. Such a superposition is illustrated in Figure 6.11, for a postulated pad that experiences the glassy-solid shoulder at about 10^4 Hz.

For a pad with this compliance response, we can observe several behaviors. First, the impulse that the wafer initiates on the pad is clearly in the viscoelastic region. The response generated by running across the pad grooves varies, depending upon the pad speed, but generally it also behaves viscoelastically. However, all impulse generators smaller than the pad grooves experience a glassy solid material response from the pad. It is interesting to note that neither the separation between the die (die street), nor the spacing between interconnect lines is differentiated by the circuit spacing, since all phenomena generate a similar pad material response. Some validation of this could be derived from studies showing that the removal rate is generally independent of line pitch, as illustrated in Figure 6.12.[5]

It is the frequency effect of viscoelastic material that makes the static measurement of polymer properties less valuable than normally perceived. More work needs to be done in characterizing the typical polymers used in CMP for their frequency responses, which will allow a more accurate positioning of their behavior. Although the glass transition point is often used as a characteristic metric for these polymers, perhaps the shoulder of the curve (marking the actual transition from the viscoelastic to the glassy solid region) would be more useful.

One aspect of the impulse response that needs to be addressed to form a more complete picture of actual pad responses is the effect of temperature. The simplest approach to transposing the response results to changes in temperature involves the formalism of time–temperature equivalence, which postulates that the response at different temperatures can be modeled by shifting the response along the time axis. If the response shift is more complicated, scaling procedures[6] can be applied. The shift factor (a_T) identifies how much of a time shift is needed to represent the required temperature change, as shown in Figure 6.13.

The procedure for obtaining the shift factor was proposed to Williams et al.[3] by Ferry[8] through constructing a series of creep plots at different temperatures and then looking to see what shift is required to reconstruct the composite creep plot (Figure 6.14).

The amount of shift needed for each temperature point can be graphed, as shown in Figure 6.15. Thus the time–temperature equivalence can be calibrated for a particular pad and used to identify the probable shift in the impulse response curve that could be expected during a typical CMP process.

Applying the shift factor plot to the impulse–response curve with typical temperatures generated during the CMP process, a reasonable estimate of the shift can be obtained. An example of such a set of shift curves is shown in Figure 6.16.

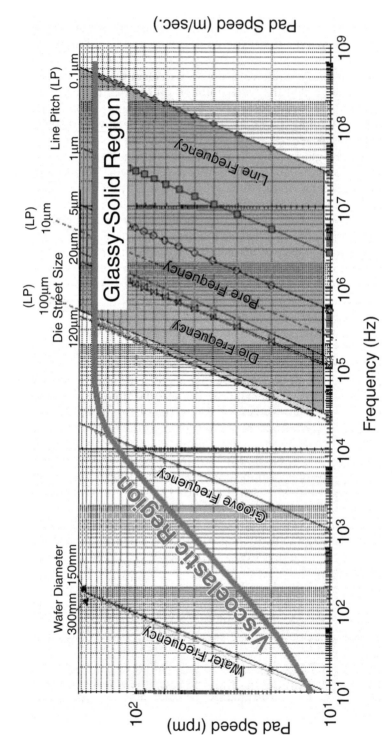

Figure 6.11 Superposition of pad creep curve on characteristic stress impulses.

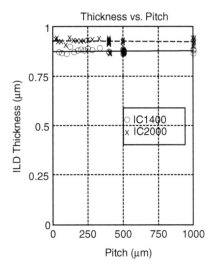

Figure 6.12 Dependence of removal rate on interconnect pitch.[5]

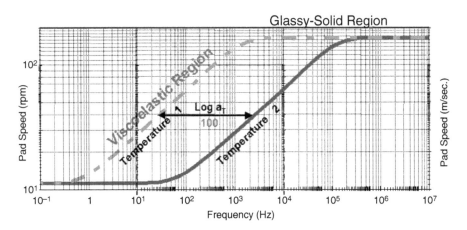

Figure 6.13 Change in material response as a result of a temperature change by using the shift factor a_T.

Of the relevant time scales that drive the frequency effects of the polymer pads, many of them have their origin in the physical structure of the polishing pad. The largest scale of these is represented by the groove. But the basic nature of many of the typical polymer pads includes a somewhat random interruption of the polymer material in the form of pores. The pores have three major impacts: (a) they modulate the abrasive surface presented to the wafer; (b) they break up the homogenous nature of the pad material itself, causing modification of the physical properties of the material; and (c) they provide microreservoirs for both slurry distribution and byproduct

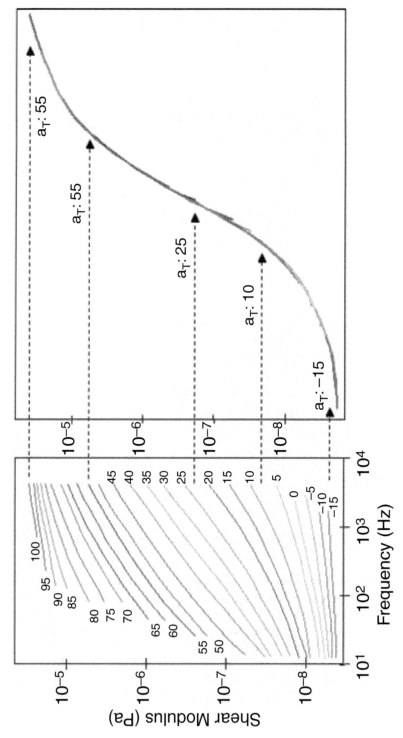

Figure 6.14 Composite shift factor curve. (After Ward[2] and Ferry.[8])

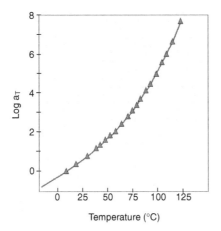

Figure 6.15 Shift factor versus temperature (After Ward[2] and Ferry.[8])

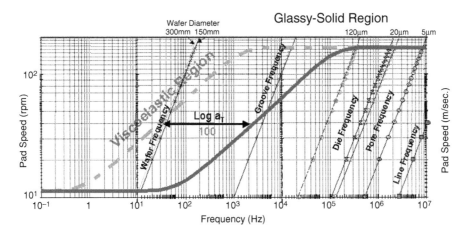

Figure 6.16 An example of shifting the impulse–response curve through temperature effects.

accumulation. Each of these three effects is significant and must be understood in some detail.

The pores are formed from bubbles during some of the typical manufacturing processes of polymer materials. The size distribution and density distribution of both are important to performance during the polishing process. Figure 6.17 shows two scanning electron micrographs of a polymer with such pore structures; the first is the sidewall cross-section of the material, and the second is the top surface after some use.

The mechanical characteristics of the polymer material are obviously modified by the existence of these pore structures. The characteristics of the polymer surface with pores can be determined by various means. For

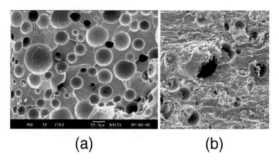

(a)   (b)

Figure 6.17 SEM of a typical polymer pad (sidewall cross-section and worn surface).

instance, the SEM can be overlaid with a grid and the physical surface structure can be measured, as shown in Figure 6.18a. The mean linear distance between pore interruptions can be measured and plotted as shown in Figure 6.18b. Measurements such as these can provide quantitative values to the pore frequency line on the characteristic stress impulse curve (such as in Figure 6.9). Different pads from the same manufacturer (or from different manufacturers) can be compared with this method for pore size and density distribution (as seen in Figure 6.19)

From such a comparison, the pore density can be factored into the measured mechanical properties of the polymer to obtain a more realistic estimate of the appropriate values. Also any effective shift in impulse response frequencies contributed by the pores can be assessed (see Figure 6.20). Although the visual comparison of the two pad cross-sections gives the impression that the right-hand sample has a tighter pore distribution, the standard deviation for the pore diameters of the two samples differed by almost 25%, while the standard deviation of the plateau lengths only differed by about 3%. This shows that the plateau length distribution, which could be considered to correspond more directly to the abrasive surface removal rate, is a combined function of both the pore density and the size distribution. Other interesting characterizations can be obtained at the same time. For instance, the exposed pore area distribution can be measured (Figure 6.21). This is useful in estimating the capacity of the pores to function as slurry or byproduct reservoirs.

From the grid measurement (Figure 6.18a) the total measured pore area adds up to about 72,000 μm^2, representing 65.5% of the pad surface area. But from the worn surface SEM (Figure 6.17), the effective abrasive surface area is visually much larger than this estimate; this is most likely due to the natural surface variations that are introduced as the pad is worn during the planarization process. These surface variations, along with the variations in effective elastic modulus locally presented to the surface by the pore distributions, creates a real-world force response that is very difficult to characterize by analytical solutions. In many mechanical domains, the time and spatial averages of these effects would be sufficient. But in CMP, the specific circuit results span such wide ranges of temporal and spatial scales that the

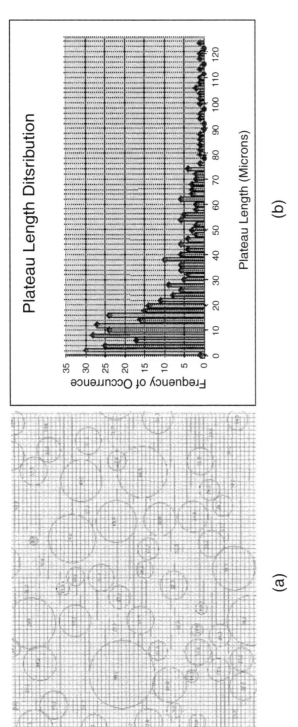

Figure 6.18 Polymer pore surface characterizations: (a) measurement grid from SEM and (b) plateau occurrence frequency distribution.

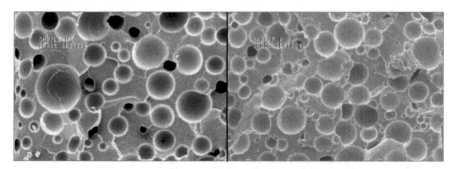

Figure 6.19 Two pad sample SEMs for pore structure comparison (the two samples have the same magnification).

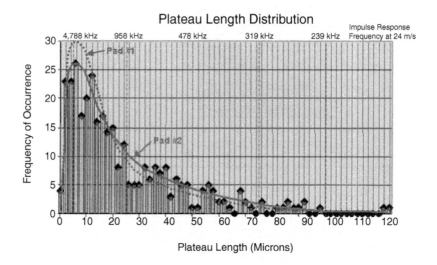

Figure 6.20 Comparison of plateau length distributions for two pad samples.

above effects are very important (refer to Figures 1.30 and 1.31). The understanding of the pore contributions to the mechanical properties of a polymer can be appreciated from the study of elastomeric foams.

6.2 Elastomeric foams

A great deal of work has been done on understanding the structure and properties of what are called *cellular solids*, principally because of their importance in structural engineering. Relevant application of this knowledge will be applied here to illustrate the theoretical background for the expected behavior of polyurethane under CMP polishing conditions.

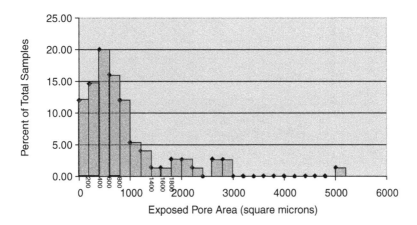

Figure 6.21 The distribution of exposed pore area at the pad surface.

The discussions often begin by partitioning materials into open-cell and closed-cell varieties (along with divisions representing isotropic and anisotropic aspects of the materials). Our discussion will revolve solely around the closed cell forms of elastic materials, which represent the most common materials in use for CMP polishing pads. Although many of the man-made closed-cell foams behave like open-cell structures because of the action of surface tension at the cell boundaries during manufacturing, the scanning electron micrographs of pad cross-sections clearly illustrate the closed-cell structure of the polishing pad made with foamed polyurethanes (Figure 6.17).

The two primary properties of these elastomeric foams are (1) the compressive modulus of their bulk material and (2) the material's relative density ρ/ρ_s (ρ_s represents the bulk material density without pores). The relative density of these materials can vary from 1 to nearly 0.01. While typical packaging foams show a relative density of 0.05, the typical pad materials have relative densities nearly an order of magnitude higher. The four basic deformation modes for cellular materials are the linear elastic mode, the nonlinear elastic mode, the plastic collapse mode, and a variety of fracture modes.[9–13] Most of the studies model the material pore structure with a hexagon-shaped pore to constitute the basic model cell. While this is convenient for some modeling systems, we must eventually appreciate any variations to the findings arising from circular pore shapes (which is the dominant pore shape for polishing pad polyurethanes).[9–11,14]

The primary analysis consists of understanding the mechanical deformation properties of such a structure under various forms of loading. This entails understanding how the physical dimensions and the material density are reflected in the effective distributed mass of the cell and its moment of

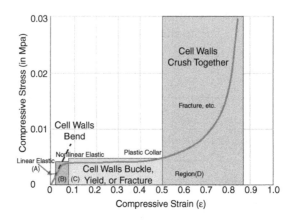

Figure 6.22 Typical stress–strain curve for foam materials.

inertia. The basic stress–strain curve is illustrated in Figure 6.22; here the four basic determination modes have been illustrated by different shaded rectangles, and are represented by separate regions in the stress–strain diagram that exhibit very different response slopes.

The basic deformation modes are represented by regions in the stress–strain diagram and exhibit very different response slopes.

6.2.1   Basic cells

Simulations of the effect of stress on closed cells are typically handled by representing the material between the cells as a simplified cell wall structure. Much can be learned about the elastic and nonelastic response by simplifying the cell structure to a basic hexagon and identifying the unit cell for repetition purposes (Figure 6.23a). The basic mechanical response will be determined by the response of the hexagonal cell structure, and then further refinements can be determined by studying a spherical cell structure (Figures 6.23b and c).

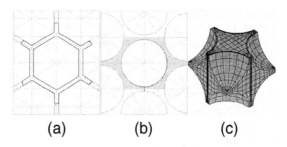

(a)   (b)   (c)

Figure 6.23 Cell structure approximations: (a) classical hexagonal cell cross-section, (b) spherical cell cross-section, and (c) spherical cell — finite element representation, at 3/4 view.

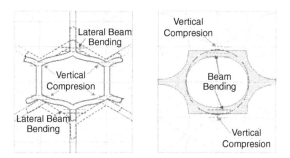

Figure 6.24 Hexagonal and spherical cell wall bending.

When the basic cells are compressed, one would expect some combination of vertical compression as well as lateral bending of the cell walls (as illustrated in Figure 6.24). Within limits, such wall bending produces an elastic (Hookean) response, characterized by region A in Figure 6.22. Although for light loads, this may be the characteristic condition of the polymeric material, it represents only a few percentage points of the available stress–strain curve response.

6.2.2 Young's modulus

Calculations from a mechanical beam bending approach leads to an expectation for the relative Young's modulus to be proportional to the square of the relative material density:[9] $E/Es = C_2 \, (\rho/\rho_s)^2$. When plotted on a log–log graph, the expectation is that the points will lie on a line with a slope of 2. Ashby[9] has plotted some 14 different experimental studies for both ceramic and polymeric foams; the open-cell foams (open symbols) are in good agreement with the theory, and the closed-cell foams are also in reasonable agreement, with slopes ranging from 1.47 to 2. As one might expect, the closed-cell foam structure responds with a smaller slope, and the relative Young's modulus moves closer to bulk polymer behavior (Figure 6.25).

Beyond the linear region in Figure 6.22, there is a characteristic behavior dominated by the elastic or plastic buckling of the supporting structural columns.

6.2.3 Cell wall buckling

When compression and/or shear rise sufficiently, the symmetrical elasticity of the unit cell collapses, and the cell walls begin to buckle (Figure 6.26). The response to increased stress no longer produces a linear strain, and the material begins to respond to nonlinear elastic behavior, characterized by region B in Figure 6.22.

Observation of octagonal cells with a force component normal to a beam of length l predicts a stress proportionality of F/l^2. From this, the plastic

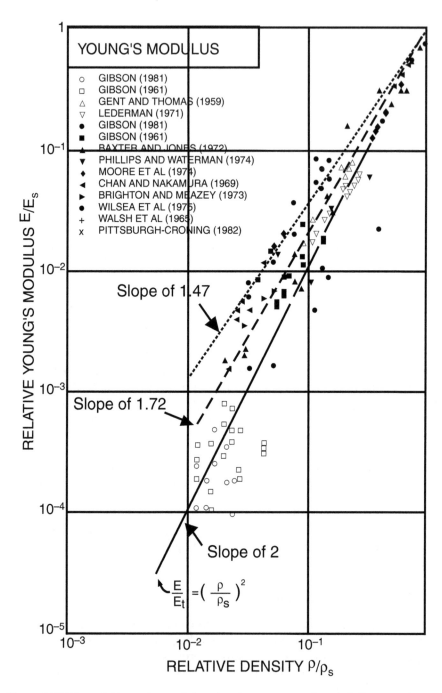

Figure 6.25 Plot of Young's modulus for foam materials of different relative densities.[9]

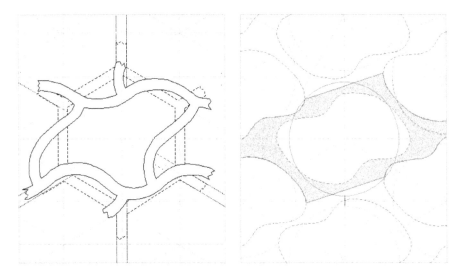

Figure 6.26 Octagonal and spherical cell wall buckling under compression.

collapse stress σ_{pl} can be shown to be proportional to M_P/l^3 (where M_P is the plastic moment of the beam structure), which point to the relationship:[9]

$$\sigma_{pl}/\sigma_y = C_4(\rho/\rho s)^{3/2} \qquad (6.33)$$

In a plot similar to Figure 6.25, the log of the relative plastic collapse stress is plotted as a function of the log of the relative density[9] for nine different experimental sets of data (Figure 6.27). There is general agreement with theory yielding a 3/2 power for the relative density relationship, although the closed cells appear to support an exponent closer to 1/2.

Since polymeric foams collapse upon compression, they do not respond to indentation by spreading laterally (displaying a Poisson's ratio of nearly 0.04). As a result, for materials with relative densities ≤ 0.3, their indentation hardness is essentially equal to the plastic collapse stress,[16] instead of the more familiar $H = 3\sigma_y$ relation for dense solids.

When the force on the polymer material exceeds the fully plastic moment of the weakest element of the cell structure, plastic collapse occurs (characterized by region C in Figure 6.22). In effect, the weakest elements become a plastic hinge and deform to create a collapsing cell that responds to nearly constant stress (σ_{pl}) with large deformations (see Figure 6.28).

While elastic buckling of structural walls is fully recoverable, plastic collapse of the plastic hinge sections is not. The theory of plastic collapse corresponds well for materials with a relative density of 0.3 or less; materials with greater relative densities do not follow theoretical predictions because their cell partitions are too thick to buckle or hinge readily. The data represented in Figure 6.18, showing SEM from 6.17a, estimates a relative density of 0.345.

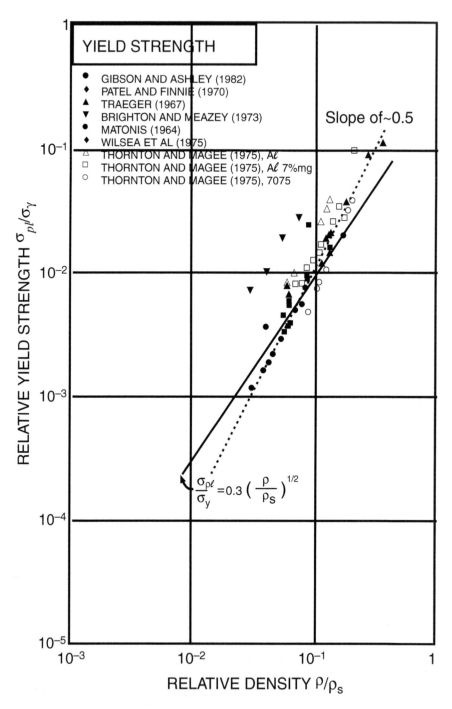

Figure 6.27 Relative plastic collapse stress for open- and closed-cell foams.[9]

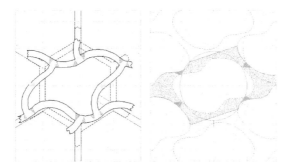

Figure 6.28 Plastic cell wall collapse for hexagonal and spherical cells under compressive shear.

Some simulation work indicates that for closed-cell structures, the plastic deformation mode occurs much sooner than elastic buckling.[13] This prediction is understandable for compressive stress, but it is less likely to hold true under the addition of shear stress (see Figure 6.29).

The three basic mechanisms for cellular solid stress–strain response are illustrated in the deformation map[9] in Figure 6.30. Lines indicating the relative density of the material under test identify which response region will dominate under the applied conditions.

Small strains are typically confined to the linear response region, after which the region of plastic collapse takes over. At high strains, the material will typically experience a densification response in which the cell walls have been fully crushed and the bulk material properties begin to be exhibited. As one would expect, the greater the relative density, the smaller the role played by the collapse of the cell walls. At somewhere near 0.3 relative density, the cells are sufficiently apart that the collapse mechanisms are insignificant (illustrated in Figure 6.31).

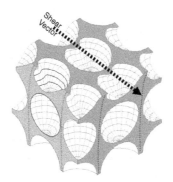

Figure 6.29 Stacked spherical cells under shear force.

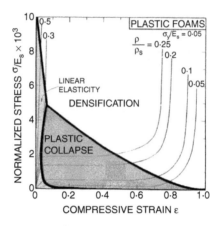

Figure 6.30 Deformation mechanism map.[9]

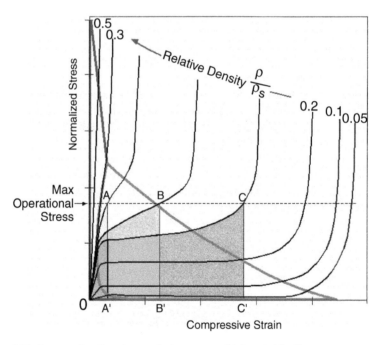

Figure 6.31 Area under the stress–strain curves. (After Ashby.[9])

The response of a polyurethane sample[17] is shown in Figure 6.32a and is superimposed on the deformation map in Figure 6.32b. Again, the material response indicates behavior appropriate for a relative density of nearly 0.3.

Whereas the discussion of crushing strength of cellular polymer material may not be very relevant to CMP pads, the surface deformations induced by the frictional shear forces argue for a discussion of the tensile fracture

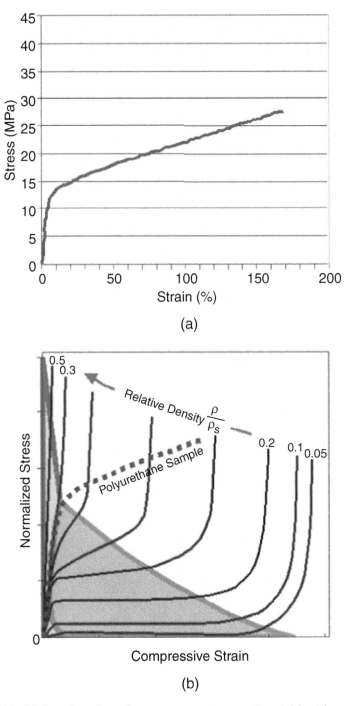

Figure 6.32 (a) Sample polyurethane stress–strain curve[17] and (b) with overlay on deformation map diagram.

strength in cellular materials. General theoretical work would indicate that compressive and tensile responses should be similar; in practice we find that tensile fractures occur much more readily. The mechanism for this is the propagation of nanoscale cracks, often seeded by material defects. Stress concentration related to the crack or material flaw dramatically increases the probability of enhanced failure.

The probability of crack propagation can be calculated for the classic hexagonal cell and can be related to cell dimensions and the applied stress.[9] A new constant is involved in the resultant equations; it is related to cell size (since crack propagation is a combination of tensile stress and bending moment). For CMP pads, the equivalent areas of concern will be the cell density (related to the relative density) and the cell size distribution (visualized in Figure 6.33). It is the cell proximity or overlap probability that will determine the material's inherent propensity to support crack propagation. Also, since much of the shear stress concentration is at the top frictional surface, the damage done to the inherent cell structure by the CMP process is likely to determine much of the material's wear response. It is probable that the shape, size, and distribution of abrasives play an important role in the material's wear characteristics.

Two probability distributions are most likely to be important: (a) the cell density probability distribution (which is the distribution of cell center locations) and (b) the cell size distribution (around the cell center locations). The combination of these distribution functions, bounded by the material relative density (say, 0.3), should result in the cell wall thickness probability distribution. A postulated cell wall thickness probability distribution can be seen in Figure 6.34a. For a given process at its typical temperature, there will be a wall thickness that will respond to the applied force within wall deflections, and one resulting in wall plastic deformations. Increasing the relative density of the pad material will shift this curve toward thicker walls, with lower probabilities of each of the above mechanisms. The effect of these mechanisms cannot be derived from the material density.

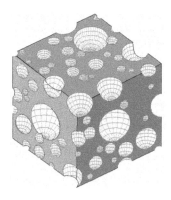

Figure 6.33 More realistic spherical pore distribution.

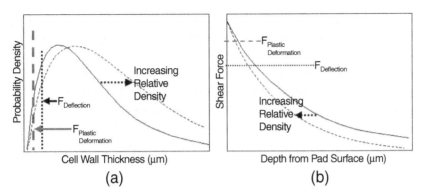

Figure 6.34 (a) Cell wall thickness probability distribution and (b) shear force depth penetration.

The effects in Figure 6.34a are represented at a particular depth into the pad. At deeper levels into the pad, the shear force diminishes. With further penetration, each mechanism reaches a threshold depth beyond which the effect is no longer viable (Figure 6.34b). It has been proposed that the shear modulus should scale similarly to the bulk modulus,[9] but an *open cell* surface behavior does not represent shear forces in the same way that *closed cell* material does. Due to the lack of structural support, the surface response should be more easily determined by the open pore structure. The material experiencing greatest strain would then be the trailing pore edges, which will be the first part of the surface plateau to withstand the shear force of the wafer surface structure. Beyond that, whatever viscoelastic response the pad has at these frequencies will allow the leading edge of the wafer surface protrusion to sink into the cavity of the pore. The trailing edge of the pore (the leading edge of the plateau) will feel the impact of suddenly being forced to ride down over the leading edge of the wafer surface structure. This generates excessive plastic deformation and wear on the pore's trailing edge as it provides the greatest shear force to the wafer surface.

The conjecture that the trailing edges of the pores suffer damage (which, we speculate, may be plastic deformation) is supported by the scanning electron micrographs of a pad before and after polishing (Figure 6.35). This is most evident in the marked pores of the second SEM, where the leading edge is intact (although sometimes folding into the pore) and the trailing edge has been broken out. As the trailing edge of the pore is worn back, the direction of the shear force vector rotates to decrease its perpendicular component (which decreases its effectiveness for material removal). To the extent that the trailing edge provides the most significant impulse to the wafer surface structure, the loss of this effect could impact the removal rate. It should also be noted that the decreased effectiveness in material removal of the pore effect is not readily addressable through pad conditioning, which cannot restore the original perpendicularity of the trailing edge pore wall;

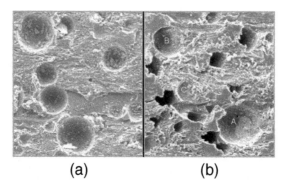

Figure 6.35 Comparison of pore trailing edge before (a) and after polishing (b).

for conditioning to be effective if this removal mode dominates, the overall pad surface would have to be removed to expose fresh pore trailing edges.

It may also be that the pore effect is a source of local nonuniformity, which is averaged through the rotation of the wafer (and pad, for rotary tools). Both linear and rotary CMP tools will always present the same trailing edge of any pore to the wafer surface, in contrast to the orbital CMP tools, which will have the opportunity to present more of the pore edge to the shear motion.

Figure 6.36a is an overlay of a pad SEM (from Figure 6.19a) in which small cell wall thicknesses have been highlighted. The frequency distributions for minimum cell wall thicknesses are plotted in Figure 6.36b. The overlap of pores (shown as zero wall thickness) appears to have its own distribution. Once a process has been established and the maximum wall thickness that participates in plastic deformation has been established, the

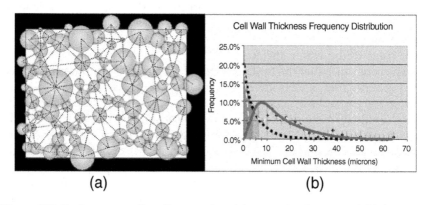

Figure 6.36 Surface pore cell wall separation: (a) separation layout and (b) frequency distribution.

probability of occurrence can be estimated. For these data, if we assume that wall thicknesses greater than 3 μm would flex rather than deform, the trailing wall deformation would probably be somewhere around 5 to 8%.

References

1. R.P. Feynman, *Lectures on Physics*, Vol. 2, Addison-Wesley, Reading, MA, 1963, chaps. 38 (Elasticity) and 39 (Elastic Materials).
2. I.M. Ward and D.W. Hadley, *An Introduction to the Mechanical Properties of Solid Polymers*, John Wiley, New York, 1993.
3. N.V. Gitis, A.J. Jin, and D.R. Craven, Tribology Case Studies for Copper Removal Optimization, Seventh Int. Conf. Chem. Mech. Planarization, AVS, 2002.
4. C. Zener, *Elasticity and Anelasticity of Metals*, Chicago University Press, Chicago, 1948.
5. D. Boning, D. Ouma, and J. Chung, Extraction of Planarization Length and Response Function in Chemical-Mechanical Polishing, MRS Symposium Q, 1998.
6. N.G. McCrum and E.L. Morris, On the relationship of the viscoelastic functions of polymeric solids at different temperatures, *Polymer*, 5, 384, 1964.
7. M.L. Williams, R.F. Landel, and J.D. Ferry, The temperature dependence of relaxation mechanisms in amorphous polymers and other glass-forming liquids, *J. Am. Chem. Soc.*, 77, 3701–3707, 1955.
8. J.D. Ferry, *Viscoelastic Properties of Polymers*, John Wiley, New York, 1961, chap. 11.
9. M.F. Ashby, The mechanical properties of cellular solids, *AIME Metallurical Transactions A*, 14A, 10, Sept. 1983.
10. L.J. Gibbon, Ph.D. thesis, Engineering Department, Cambridge University, 1981.
11. L.J. Gibson, K.E. Easterling, and M.F. Ashby, The structure and mechanics of cork, *Proc. R. Soc. London*, A377, 99–117, 1981.
12. M.R. Patel and I. Finnie, Structural features and mechanical properties of rigid cellular plastics, *J. Mater.*, 5, 909, 1970.
13. J. Oung, J.H. Lee, and H. Liang, Modeling of the mechanical behavior of polyurethane CMP pads, *Int. J. CMP ULSI Multilevel Interconnection*, 1, 1, 85–96, Spring 2000.
14. L.J. Gibson, M.F. Ashby, G.N. Karam, U. Wegst, and H.R. Shercliff. The mechanical properties of natural materials. II. Microstructures for mechanical efficiency, *Proc. R. Soc. London*, A450, 141–162, 1995.
15. F.K. Abd. El Sayed, R. Jones, and I.W. Burgess, A theoretical approach to the deformation of honeycomb based composite materials, *Composites*, 10, 209, 1979.
16. M. Wilsea, K.L. Johnson, M.F. Ashby, Identification of foamed plastics, *Int. J. Mech. Sci.*, 17, 457, 1178, 1975.
17. Dr. Jibing Lin, private communication, 2003.
18. D. Stein, D. Hetherington, M. Dugger, and T. Stout, optical interferometry for surface measurements of CMP pads, *J. Electronic Mater.*, 25(10), 1623–1627, 1996.

19. A.S. Lawing, Pad asperity structure and pad conditioning in CMP, CAMP 2003 Conf., 2003.
20. S. Lee and D. Dornfeld, Small Feature Reproducibility Workshop, University of California, Berkeley, 2003.
21. D. Hetherington, CMP, International Interconnect Technology Conference, p. 47, June 2001.
22. S.A. Lawing, Polishing Rate, Pad Surface Morphology and Pad Conditioning in Oxide Chemical Mechanical Polishing, ECS 2002.
23. S.A. Lawing, Pad Conditioning and Pad Surface Characterization in Oxide Chemical Mechanical Polishing, MRS 2002.
24. Zygo Corporation, Middlefield, CT, www.zygo.com, 2002.
25. M.R. Oliver, R.E. Schmidt, and M. Robinson, CMP Pad Surface Roughness and CMP Removal Rate, Intel internal study, unpublished.
26. N.V. Gitis, Tribology Issues in CMP, Semiconductor FabTech, 18th ed.
27. L. Doyen and D. Vacher, Analyzing Large Particles in CMP Slurry, Semiconductor International, 8/02.

chapter seven

CMP pads

Polishing pads play a central role in CMP. This chapter will provide an introduction to polishing pads following their condition process.

7.1 Pad surface characteristics

Conventional wisdom ascribes the bulk of the material removal to asperity/abrasive interaction with the wafer surface, so it is essential that the surface profiles of the polishing pad be well characterized. The classic model of the pad is that its abrasive surface is covered with polymer protrusions of differing lengths, heights, and widths. These then become the primary elements of (and are ultimately responsible for) the transmission of force to the wafer surface from the downforce pressure on the carrier head and the response pressure of the polishing pad. This picture is particularly true for abrasive-free slurry processes. However, in the conventional abrasive slurries, the force will be transmitted to the abrasive particles (perhaps shared by nonabrasive coated polymer surfaces), which then make direct contact with the wafer surface, causing material removal. Perspective on the relative sizes of the pad's surface elements can be seen in Figure 7.1a, and Figure 7.2.

We need to understand the force-bearing characteristics of the asperity-rich pad surface. The scale of this perspective shrinks another order of magnitude from that used to examine the pad's pore structures. As seen in Figure 6.16, the wafer compression action on the polymer material is likely to show a viscoelastic response (the compression span of the traveling wafer is sufficiently long for the polymer material to respond viscoelastically). The asperities in contact with the wafer surface will then suffer a viscoelastic modification. The nature of this modification will be defined by the forces at work on the asperity profiles.

Various approaches have been used to estimate the effect of the wafer downforce on the asperities,[1,2,3] including the use of elastic force response and volumetric displacement. Whereas the use of SEMs and other high-magnification metrology tools to image surface contours is a big step in understanding what the *initial* surface characteristics are, one must be prepared

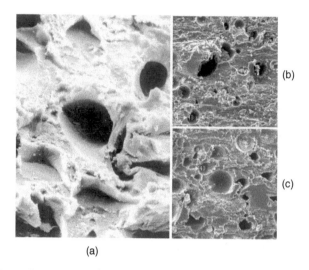

(a)

Figure 7.1 SEM of typical polishing pad (a) enlarged, (b) before polish, and (c) after sustained polish.

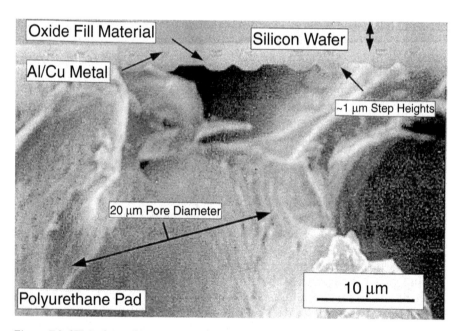

Figure 7.2 SEM of a pad.

to project some form of surface transformation to understand what the surface features are likely to be when they are under compression and shear forces (see Figure 7.3). This *shear transformation* will render an effective surface roughness based on a number of surface parameters. Under the action

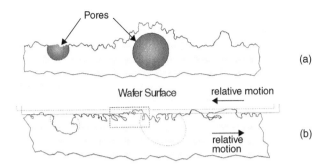

Figure 7.3 (a) Pad cross-section (from Figure 7.1a) and (b) under compression and shear.

of compressive and shear forces, the viscoelastic asperities will be drawn out in the trailing direction of the pad motion. The polymer area exposed to the wafer surface will be represented more by the asperity sidewall than by its tips. Thus the response force will be more equivalent to the asperity thickness than to its height. But as the asperity is laid sideways by the action of the shear force, the response force will also be a function of the pad surface profile into which the asperity is being compressed. It is likely that a new surface roughness parameter needs to be developed that can map more easily into the shear surface response force. An artist's rendering of such a surface profile transformation is seen in Figure 7.3 (rendered from Figure 7.1a). The dashed rectangle in Figure 7.3b is enlarged to illustrate the response forces in Figure 7.4.

Once the asperities have been viscoelastically distorted by the compression and shear forces of the wafer, the leading edge of the asperities will then be subjected to the high-frequency impulse forces of the wafer surface features. As was explained in Section 6.1.4, the high-frequency impulses are most likely to experience the pad as a glassy solid and thus generate plastic deformation at the *transformed* leading edge portion of the asperities (the arrows in Figure 7.4). As the leading edge suffers repeated plastic deformation, the probability of shearing off sections of the asperities increases. Lawing[4] has estimated an effective spring constant for the IC1000 pad of

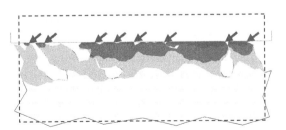

Figure 7.4 Compressed surface asperities (enlarged) with force response illustrated by shading.

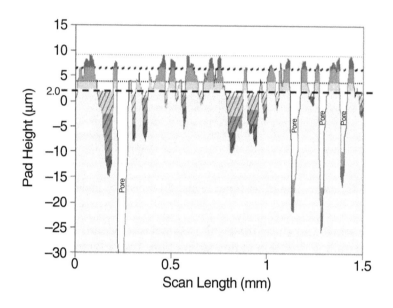

Figure 7.5 Volumetric displacement. (After Lawing.[4])

roughly 4.5×10^8 Pa. While the elastic spring constant is important in reaching a force-balance condition for asperity compression, the approach becomes much more complicated when a shear/compression scenario is examined. Under shear forces, the asperities are folded over, stretched out, and compressed into the adjacent surface topography (Figures 7.3b and 7.4). Thus a rigorous analysis would have to combine the probability of asperity height with the probability of adjacent asperities or depressions.

Another approach that does not suffer from such complexity is an analysis of volumetric displacement (Figure 7.5).[5] First one would have to subtract the pores to get a more accurate asperity structure assessment. Although neighboring asperities could fold into the void created by a pore, much of the pore would remain unfilled by this folding action. The pore removal requires some algorithm to identify pores by their surface characteristics. We have seen that pores can have diameters down to 10 μm, at which point they become difficult to differentiate from asperities. If one were to classify pores arbitrarily as structures with negative topographical excursions of 20 μm or more, then the remaining surface features could be realistically taken as asperities. A complete folding would require that the available plateau length be equivalent to the folded asperity height. A uniform flattening of the asperities would require that the surface frequency (frequency of asperity perturbations from the nominal pad plane) be related to the asperity aspect ratio.

Once the planar level is established with all the asperity volumes filling all the volumetric cavities (with the exception of the pores), then a

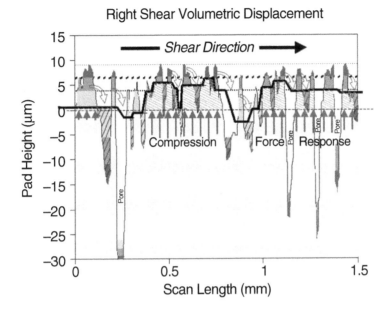

Figure 7.6 Modified shear volumetric displacement. (After Lawing.[4])

compression level can be determined from the compression constant of the material and the downforce pressure.

Because the asperities are fixed on the surface and can only influence their neighboring regions, we suggest a revision. A modified (shear) volumetric displacement, where the asperity volume is displaced in one direction (the shear direction), is probably more realistic. Figure 7.6 illustrates the differences generated with the shear volumetric displacement. Again, the pores are left relatively alone since they will not be filled to any great extent. But the asperity volume is folded only to the right, and the adjacent cavity is filled until it is flat with the asperity level. This generates a nonuniform topography (the solid line), which puts locally high spots on the pad surface under compression (represented by the areas with jagged diagonal lines). A nonuniform compressive force response is generated across the pad surface, which will be a most significant factor for material removal.

Precision interferometric optical metrology systems are commercially available[6] to render three-dimensional surface profiles to micron and submicron precision. A profile scan of a pad surface[23] is illustrated in Figure 7.7, with the relatively flat surface areas shown by the areas with jagged diagonal lines. As we can see from previous figures, scanning electron micrographs can provide much finer resolution. It is very difficult, however, to translate the SEM image into a numerical profile. Such capabilities of optical interferometric tools allow a quantitative description of the pad surface at the micron level, enabling comparative studies under changing conditions. These studies will

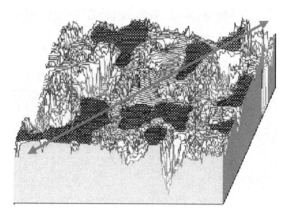

Figure 7.7 Interferometric profile of pad surface.

address such key issues as the relationship of bearing area to material removal characteristics.

One of the important questions is why asperities stop working as the planarization process progresses. Three key roles of the asperities, we suggest, are (1) a force transmission mechanism (between the downforce pressure and the wafer surface), (2) a participation in the slurry distribution to the wafer surface, and (3) an agent in the removal of the polishing byproducts. So for a degradation in material removal rate with polishing time, the asperities must be losing functionality in at least one of these areas.

It is difficult to see how the force transmission would degrade as the asperity topography flattens out with wear. To a first-order approximation, the volume of material removed should be linearly proportional to *the force times the bearing area* ($F \times A$). As the asperity peaks are smoothed out to a planar surface, the individual forces that were exerted by the fewer peaks is reduced, but in the same proportion as the increase in the bearing area. Unless one can justify a threshold force (vis-à-vis activation energy), linearity should hold and the material removal rate (MRR) should not decrease as the individual asperities are worn down.

The slurry distribution issue is more complicated, if only because there are both mechanical and chemical effects involved. The mechanical component may be tied to the slurry viscosity as well as all the mechanical considerations revolving around the slurry abrasives. The chemical component will be affected by a host of conditions, such as stagnant boundary layers, reactant concentrations, local pH, and isopotential conditions, and so on.

7.2 Slurry depletion

In either case, an important study would be to address the depletion effect of the slurries being used for CMP planarization. This would answer the question of how far into the pad surface effective material removal is

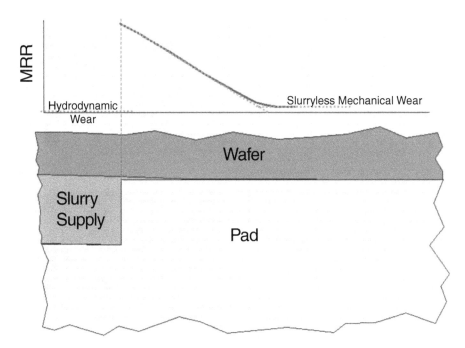

Figure 7.8 Effective slurry depletion.

accomplished without rewetting the wafer surface with new slurry. One could imagine a smooth pad surface without a slurry coating, adjacent to a groove completely filled with slurry, and the wafer surface was then wiped across it. Assuming the groove coats the wafer surface uniformly with fresh slurry, what is the effective removal rate as a function of distance from the groove (Figure 7.8)?

Figure 7.9 illustrates the asperity height distribution for a pad with two different processing regimes. The top set of graphs corresponds to a pad under process conditions that maintain a uniform surface roughness (conditioning dominated). The bottom set of graphs corresponds to process conditions in which there has been significant deformation of the asperities under the combined action of the slurry and the wafer (wafer dominated).[7]

A real-time measure of the coefficient of friction is essential to a well-controlled CMP process. It is possible that this could be done by a coefficient of friction (COF) probe, mounted on the CMP tool, which could feed back observed values to the process control computer. There are several requirements in making a measurement that truly reflects the process conditions, the first of which is that the probe surface material should be the same as the wafer surface material (so the COF you read reflects what the wafer experiences). That situation, of course, is already inherent in the wafer carrier head, but the signal-to-noise ratio in trying to read the COF off the wafer carrier head is often too great to be of use.

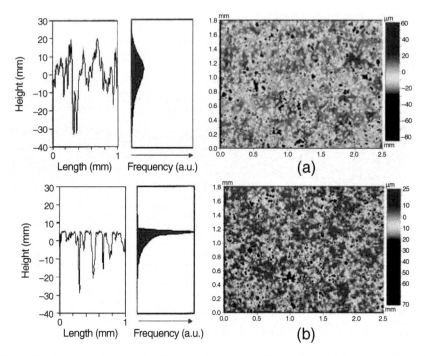

Figure 7.9 Asperity height distribution for (a) conditioned and (b) unconditioned polished pad.

This would require the probe head to be of oxide when processing oxide, or copper when processing copper. The probe head would need to be placed in a representative part of the pad to intercept the slurry stream similar to what the wafer intercepts. Since the probe would glaze the pad in a similar fashion to the wafer, the probe head should probably be as small as is feasible. It will also need to be testing a section of the pad that gets a representative conditioning similar to that intercepted by the wafer. It should be circular, so that the ratio of the leading edge length to the disk surface area is in the same proportion as the wafer head. Ideally, the probe head would rotate at the same angular speed as that of the wafer carrier head. For linear tools, the probe should probably be positioned on the center line of the pad; for rotary tools, the probe would ideally oscillate so that it is traced out the center line of the wafer position. The surface profile associated with scan length is illustrated in Figure 7.10.

The mechanical difficulties in reading the COF directly for a process may not be commercially feasible, and yet much can be gained by relative measures of COF during a process. Commercial probes are available that can be mounted on the CMP tool and provide a direct readout of the coefficient of friction between the slurry/pad system and the probe head. If the probe head is made from a material that can be used as a good standard for most processes (say sapphire), then relative changes to the COF can be monitored

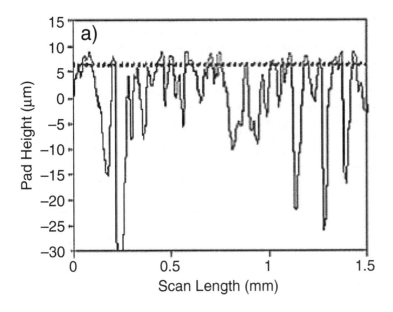

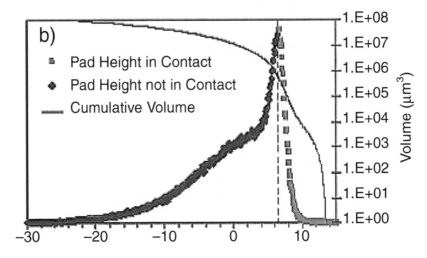

Figure 7.10 Surface asperity distribution and fit.[5]

or even integrated into the tool to provide *in situ* process control. Somewhat similar to COF is the ability to sense the surface interaction through a process known as acoustic emission (AE), as shown by the lower line in Figure 7.11. Monitoring AE can also identify changes in surface interaction in ways that can be used to support process control.[2]

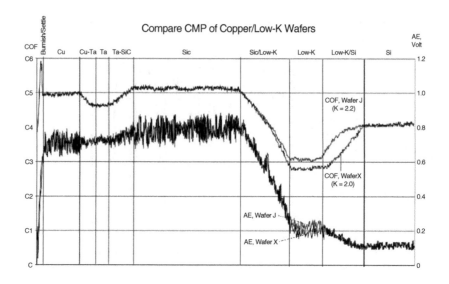

Figure 7.11 Using COF and AE to identify layer transition points for two different wafers.

There is some discussion that the primary degradation of MRR is due less to pad asperity deformation than to byproduct accumulation. The optimum approach to maintaining consistent MRR could be quite different depending on the true nature of the mechanism (physical or chemical). For instance, handling a byproduct accumulation may be addressable chemically, which would reduce the wear on the pad (also increasing its lifetime). More work should be be done to separate the MRR performance based on the extent of polymer roughness or byproduct presence.

7.3   Pad conditioning

Pad conditioning is a process that attempts to provide three important conditions: (1) maintaining a uniform roughness of the pad surface, (2) maintaining surface conditions conducive to steady-state slurry distribution, and (3) participating in the byproduct removal from the pad surface.

As we have seen, the primary functions of the pad are (a) force transmission, (b) slurry distribution, and (c) byproduct removal. The MRR is a function of all three elements, in a proportion depending upon the application, the consumable set, the device structures under process, the CMP tool configuration, the length of the process (on one wafer, or between calibrations), and so on. More study is needed to differentiate the individual effects and the interaction effects of these elements.

7.3.1 The conditioning process

The work of conditioning is sometimes computed as the product of the distance the conditioner travels along the pad and the downforce pressure on the conditioner disk. Since work is defined as distance traveled against a force in the direction of travel, a more accurate formulation would be the distance traveled times the frictional force exerted by the conditioner disk (rather than the disk downforce pressure). Any variation of the coefficient of friction (owing to changes in surface roughness, slurry lubrication layers, changes in disk diamond biting angles, etc.) would introduce errors into the calculated work using the disk downforce. If we assume that the coefficient of friction is constant (perhaps during one phase of a process step), then the work of conditioning would be directly proportional to the distance the conditioner traveled during its engagement with the pad. In a linear system, the associated calculation is fairly straightforward. But if we assume a Prestonian model for the conditioning action, as typically assumed for the wafer, then an additional velocity term must be taken into account. In a rotary system, this effect begins to play a more important role since different parts of the conditioning disk are engaging the pad at different speeds (owing to the differences in their radii to the pad center). Just as in the case of the wafer, this effect can be locally compensated for by rotating the conditioner disk (at the angular rotation rate of the pad). Again, like the wafer, the linear velocity (which is what enters into the equations) will vary depending upon the operating distance to the pad center. Although rotation of the conditioner disk can make the local area under the disk experience a uniform velocity, the moment it moves to a different pad radius, both the wafer and the conditioner disk rotation rates would need to change to maintain the linear velocity it had before. Although this is feasible, it is not an industrial practice.

The two important measurement criteria for conditioning are surface roughness and pad wear. The agent for producing these conditions is the diamond disk, with its supporting pressure and movement mechanisms. A variety of methods producing conditioner disk downforce are employed by various CMP tools, but the primary concern is to have some way of providing pressure on the disk and a calibrated feedback system for maintaining its accuracy. To provide disk position, all tools give some form of calibrated lateral motion, and often some form of disk rotation is included. For a linear system, disk rotation or indexing can be used to periodically present a new facial orientation of the diamonds (to prevent buildup from degrading the cutting action). For rotary systems, disk rotation can also be used to even out velocity mismatches between the inner and outer edges of the disk. Figure 7.12 shows the effectiveness of conditioning. The need for conditioning to provide a stable MRR performance has been known for a long time.[8-11]

The definition of a suitable conditioning process has outputs of MRR (both value and uniformity), pad wear, and conditioner lifetime (including the probability of diamond dropout). The input parameters are grouped by

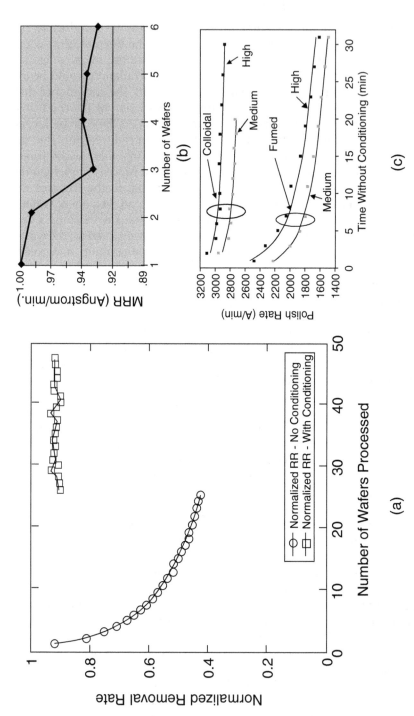

Figure 7.12 (a) Effect of pad glazing and conditioning on polish rate.[9] (b) (After Ali and Roy.[10]) (c) Rate degradation without conditioning.[11]

(a) vendor approach (which determines diamond profile, diamond density, and diamond-mounting surface adhesion) and (b) CMP tool recipe (which includes time spent at any particular pad position, conditioning downforce, and approach to conditioner surface cleaning).

7.3.2 Rate stability

One of the parameters in optimizing the conditioning process is the stability of the MRR (including both rate and uniformity). The purpose of conditioning is to provide a *consistent* surface roughness, but rate changes during the planarization process can make a consistent manufacturing process very difficult to establish. Figure 7.13 illustrates the time stability variations associated with just the choice of conditioner types, for both MRR value and uniformity.[4,12,13]

In tuning a conditioning process, the readily adjustable "process knobs" are the tool-component parameters; these include the recipe profile (the time spent at any particular pad position) and the conditioning downforce (as well as the choice of surface cleaning for the conditioner disk). An example of a tuning exploration is illustrated in Figure 7.14. Here the MRR and uniformity dependence on changes in conditioner downforce pressure are examined across a set of different conditioner disks for one set of CMP conditions.[12] We observe that the MRR value changes with downforce can be seemingly unrelated to the uniformity change with downforce. It is important to observe the effects of a parameter change on both the value and the uniformity of the MRR (since it is often difficult to predict even the direction of the response, let alone the value). In the observations illustrated by Figure 7.14, one of the conditioners studied experiences a marked falloff in MRR value as the downforce reaches 1 psi; and yet two of the conditioners drop to reasonable uniformity values (99% uniformity = 1% nonuniformity) under 1 psi downforce, while the other two do not approach that level of uniformity until twice the pressure (or more) is applied.

Some of these results are in marked contrast to other observations that indicate an increase in MRR value with increased conditioner downforce.[12,13] In this study the MRR was examined against variations in conditioner downforce, slurry abrasive type (fumed or colloidal), conditioning approach (*in situ* or *ex situ*), and pad slurry distribution support geometry (perforated or grooved). Results indicated a strong positive correlation in nearly all cases between conditioner downforce and MRR (Figure 7.15).

In the final analysis, it is the pad–slurry system's performance (under the application and tool conditions) that determines the optimum processing condition. For interlayer dielectric (ILD) oxides, this will often be measured by the removal rate value and uniformity. For other processes, additional metrics, such as dishing and erosion, must be examined for optimization. The evaluation of these additional qualifiers is more complex, because circuit pattern density plays a major role. Often the judgment will include an examination of dishing across a spread of line widths, and an examination of erosion across a spread

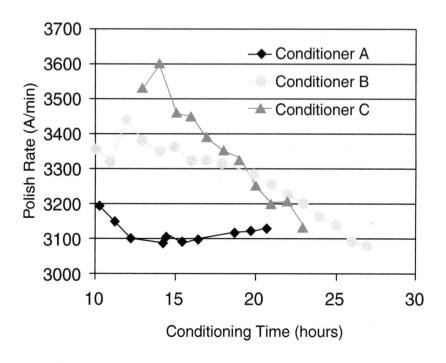

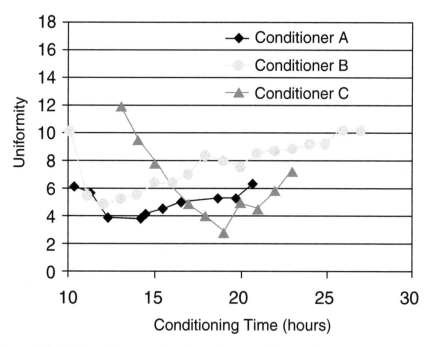

Figure 7.13 MRR stability over time for various conditioners.[4]

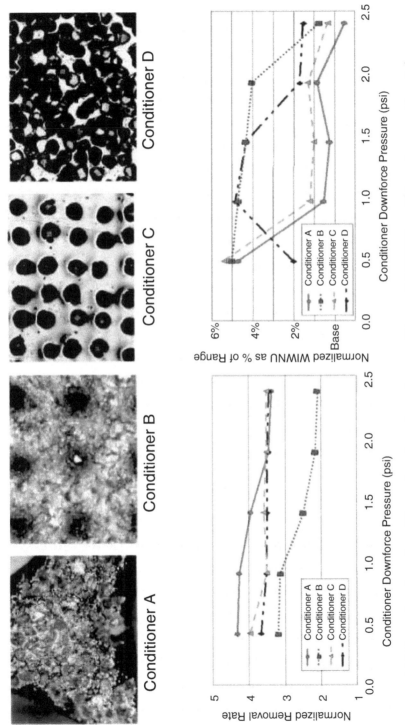

Figure 7.14 MRR as a function of conditioner downforce for various conditioners.[14] (WIWNU = within wafer nonuniformity).

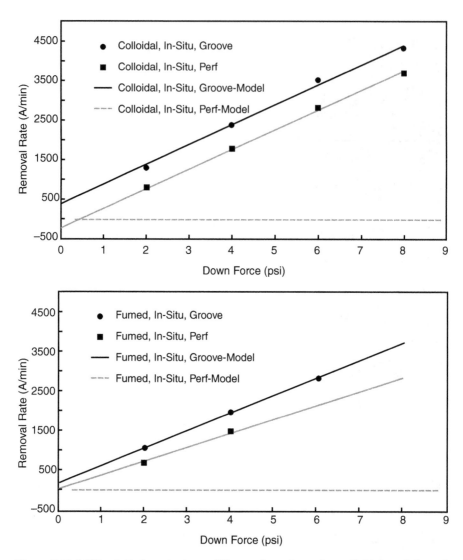

Figure 7.15 MRR plotted against conditioner downforce for colloidal and fumed silica abrasive.[13]

of pattern densities. In Figure 7.16 we see an evaluation of three different conditioning disks for dishing and erosion performance.

It would be useful to define an easily measurable metric for which the performance of the individual elements could be evaluated; then, knowing the resulting value, the local CMP factors' performance could be calibrated their projected performance. One suggested metric for pad conditioner abrasives is the relative abrasive sharpness (RAS) measurement.[15] A constant candidate for such a metric is the COF. There are readily measurable derived effects, such as the rotary platen or the linear belt motor current, or even the

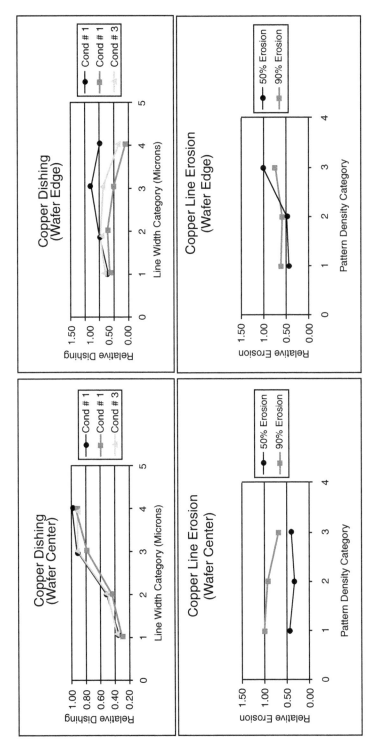

Figure 7.16 Dishing and erosion performance for various conditioners.

wafer head rotation motor current. There are also adjunct systems that can be added to the CMP tool to deliver measurable COF values for real-time process control.[16]

The final result of the conditioning process is not only a factor of the choice of conditioning disk, it is also affected by the choice of slurry abrasive. One can conceivably attribute some of this to the differences in wear of the pad asperities that different abrasive types perform. It should be noted that fundamentally different physical forms of abrasives can be used in different slurries. The different physical profiles of these abrasives can create dramatic differences in MRR characteristics. But beyond that, there is an interaction between the choice of the slurry abrasive and the conditioning effects.

Figure 7.17 identifies the MRR value as a function of conditioning down-forces from different conditioning disks acting on slurries with different abrasives.[4] The saturation of the MRR with increased conditioning pressure is an interesting phenomenon. We have now seen that variations in local conditions can produce a decrease, an increase, or a saturation of the MRR value with increasing conditioner downforce. This is a good illustration of the complexity of the CMP process; diametrically different results can be valid under different local conditions, defined by such process elements as device structure, application, abrasive, and pad–slurry system choices.

Some process designs try to use conditioning as an independent process variable, attempting to compensate for variations in other process conditions by adjusting the aggressiveness of the conditioning steps. Caution must be used, if that is the intent, to ensure that the effects of the conditioning recipe (on MRR, pad wear, etc.) have been well identified across all of the interactive

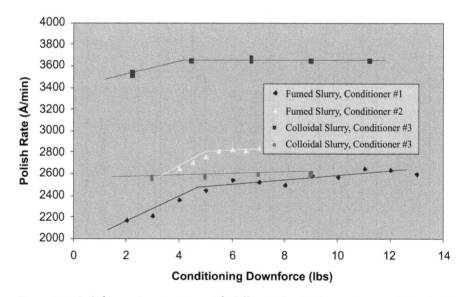

Figure 7.17 Polishing rate saturation with different slurries for various conditioners.[10]

and time-dependent process conditions. Without a well-defined response to the conditioning recipe steps across all of the various process conditions, such an adjustment would only serve to add an uncalibrated shift to an already highly interactive system.

In 1997, Steigerwald et al.[9] indicated that CMP technology was more of an art than a science. Much work is being done to understand the impact of these choices, and we continue to understand more and more about the interactive effects in the CMP process; however we still seem to be a long way from a unifying theory.

Much discussion in conditioning revolves around the issue of optimized conditioning patterns. One of the constant topics is the different action of the wafer under polish on different parts of the pad surface. For instance, if we examine the planarizing interaction with the pad from the point of view of "time-under-pressure," we would expect the pad to be glazed more strongly along the cord through the center of the wafer than along the cords near the edges (see Figure 7.18). This effect is similar in rotary and linear systems: the path of the center of the wafer contacts the pad much longer than the paths of the wafer near the edges. The effect on the wafer can be partially distributed across it for greater uniformity by rotating the wafer. However, it is clear that wafer rotation does not even out the nonuniform glazing that the pad receives.

The first-order solution is to implement more conditioning in the section of the pad that experiences the wafer center paths. This need has generated the concept of zonal conditioning in different CMP tools; the conditioning disk is forced to spend time at a pad location in proportion to the length of the wafer cord that intersects that pad location. On the surface, this concept

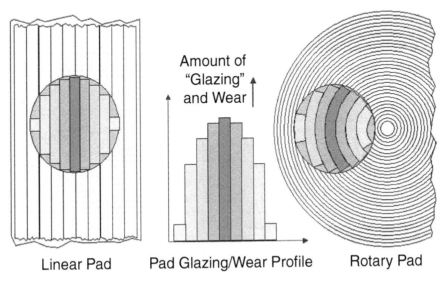

Figure 7.18 Linear and rotary pad glazing/wear profiles.

is sound; however, the implications of such a program create additional complexity. The first assumption is that we produce roughness on the pad surface proportional to time spent with the conditioner disk. In a linear system, this approximation is reasonable since the shear velocity is everywhere the same for a nonrotating disk. But in the rotary system, the shear velocity is dependent on the radial position from the center of the platen. This is apparent when viewed from the work-of-conditioning point of view. Efforts to optimize the "work of conditioning" on a rotary system have been reviewed.[11] The velocity mismatch due to the rotation of the platen is by no means an insurmountable barrier. The velocity mismatch across the conditioner disk can be compensated for by rotating the disk at rates reflecting the platen's angular velocity. However, the shear velocity of the disk will always be dependent on its radial position from the platen center. If conditioning is done *ex situ*, the platen velocity can be varied to compensate for the increased shear velocity as the conditioner moves toward the edge. If the conditioning is being done *in situ* (while the wafer is being polished), then this compensation is not practical. However, one must be careful not to confuse wear with surface roughness. Uniform wear is not the same as uniform surface roughness; the wafer cord effect would require differential wear (center heavy) to compensate for the differential glazing in order to obtain a uniform surface roughness. The true objective, obviously, is uniform (or controlled) MRR.

A further complication of a practical nature arises out of the second property of conditioning, which is the generation of pad wear. Assuming that different parts of the pad will require different amounts of conditioning to maintain a uniform surface roughness, this will lead to different amounts of wear on the pad. Thus the effect of conditioning wear is to change the contour of the pad surface from relatively flat to center-dished (for linear tools) or bimodally dished (for rotary tools). Wearing the pad unevenly will then result in changes to the downforce (pressure) distribution across the wafer (Figure 7.19).[13] This is clearly a nontrivial problem, since a great deal

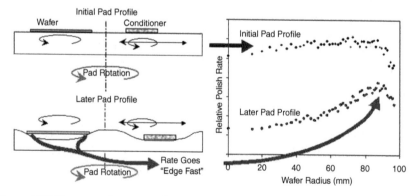

Figure 7.19 Rotary pad wear with continued polishing and conditioning.[13]

of effort continually goes into gaining a high degree of MRR uniformity across the wafer. Many of the same mechanisms used to gain uniform surface roughness create uneven wear, leading to MRR nonuniformity. The art of generating a robust industrial process can be seen in the balances that must be struck between competing effects such as these.

Furthermore, we must remember that these effects are dynamic and thus continue to change as the process goes on. One could envision a conditioning recipe that used a center-heavy zonal pattern *in situ*, and a center-light zonal pattern *ex situ*; and the balance between these effects could be proportional to the number of wafers run since the pad was installed.

Ultimately, this effect might drive the pad out of specification (either from rapid wear or from nonconformity to MRR targets) and become a fundamental source of consumable costs. Only compensating solutions that are or can be programmed to change with time (based on number of wafers processed) will be able to extend the useful life of the pad.

One interesting observation is that the basic nature of an air bearing system (common to linear tools) has some capacity to compensate automatically for this effect since the air bearing provides a natural response force to balance the head downforce pressure. If there is sufficient flexibility in the pad support medium, a pad surface that has been worn thinner will be flexed from the back until a compensating force is reached (again bringing equal pressure to bear on the wafer's surfaces).

Only when the conditioner motion is mapped onto the surface of the moving pad can the conditioner coverage uniformity (or lack thereof) be recognized. Typically the conditioner arm executes either a linear motion or a radial arc motion from side to side across the pad. Because the action of the conditioner disk on the pad is a straightforward geometrical application, it is relatively simple to create a simulation to view the effects of different conditioning patterns on the pad.[17]

The primary value of such a simulation is to identify parameters that generate what can be called effective lissajous conditions (ELC), in which the conditioner disk returns to the same position on the pad with each cycle repetition. It is important that these ELCs be identified and avoided if the pad is to be uniformly conditioned. Unfortunately, even minor changes to the repetition pattern can create major shifts in the conditioner coverage uniformity. As the conditioner finishes its traverse course across the pad, there is typically a reversal time inherent in the mechanical response to the change of directions. It is important that any inherent delays such as these be incorporated into the simulation; for instance, a reversal time of only half a second can translate to nearly 40 inches at a linear velocity of 400 ft/min, which can shift the pattern dramatically. Simulation results for both linear and rotary systems are illustrated in Figure 7.20. In the flanking figures, a reasonably uniform pad coverage is obtained at the identified parameters. Small variations, however, can result in ELCs as seen in the central graphs in this figure. The top row corresponds to a linear pad behavior, while the bottom row corresponds to a rotary pad behavior. The difference between

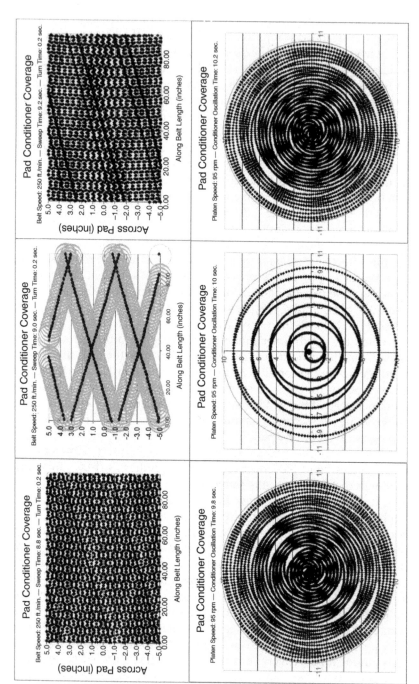

Figure 7.20 Pad conditioner coverage simulations for linear and rotary pads (showing effective lissajous conditions).

the successive horizontal figures results from a shift in conditioner sweep time of only 0.2 sec.

Zonal conditioning is not an answer to the inefficient pad coverage of ELC on either linear or rotary tools. The ELCs have to do with matching the conditioner sweep frequency to a fundamental harmonic of the pad repetition rate. Zonal conditioning changes the form of pad coverage during the sweep but has its own repetition frequency, for which the entire zonal cycle is repeated. If this repetition frequency matches a harmonic of the pad repetition rate, then an ELC will exist for the zonal pattern. As an example, Figure 7.21a illustrates an ELC for a linear pad with the zonal pattern shown in Figure 7.21b.

7.3.3 Conditioner disk — diamonds

The purpose of the conditioning disk is to present the conditioning abrasive to the pad surface in a way that will deliver the necessary pad conditioning effects. Currently, all three of the required effects are treated as a single effect and handled through physical abrasion of the pad surface with a diamond-studded end-effecor (called a disk). The use of diamonds is necessary because of the extreme hardness of the abrasives used in the slurry. The presence of such abrasives on the pad would rapidly wear down materials softer than diamond. Several major suppliers of diamond conditioner disks are available, each one with its own combination of disk properties. A variety of physical disk formats is available, from solid disks to open patterns (such as ribbed or honeycomb shapes). Likewise a variety of ways to bind the diamonds to the disk exist, and they impact disk lifetimes and possibly wafer scratch damage. There is also a choice of diamond types, which affects the shape of the abrasive as it protrudes from the disk surface.

All these factors fall into two classes: those that affect the conditioning action, and those that affect the lifetime and reliability of the conditioning disk. The first class will be the major concern here; the adhesion of the diamond to the mounting is the only effect from the second class that will be dealt with in this section.

The conditioning effect is primarily driven by two disk components, (diamond profile and diamond density), and by two tool conditioner components (downforce and shear velocity). The diamond profile is made up of the combination of the diamond shape and its protrusion from the mounting surface. The type of diamonds chosen by the manufacturer primarily determines the shape of the diamond edges presented to the pad. One manufacturer has classified these into four basic types: angular, mosaic, blocky, and cubocta (Figure 7.22).[13]

The angular diamond type is derived from natural diamonds and is the most aggressive (and least stable) of the varieties. The other three types are all derived from synthetic diamonds, and they fill out the range to the least aggressive (and most stable).

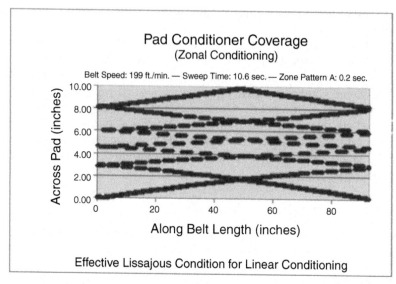

(a)

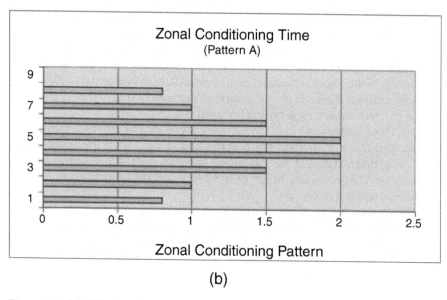

(b)

Figure 7.21 (a) Effective lissajous condition for linear conditioning and (b) zonal conditioning pattern.

A few different diamond distribution patterns are available, principally structured and random. The manufacturing of the structured type places the diamonds in an orderly array across the conditioner disk, compared to a random implementation for the alternative (Figure 7.23). On the surface, we

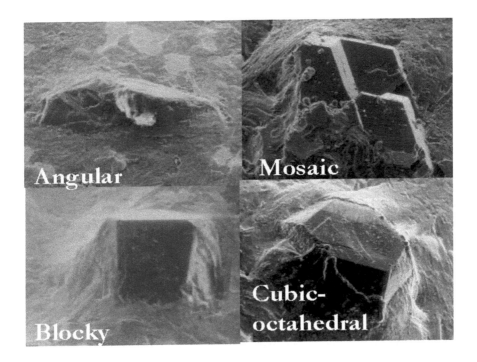

Conditioning Diamond Classification

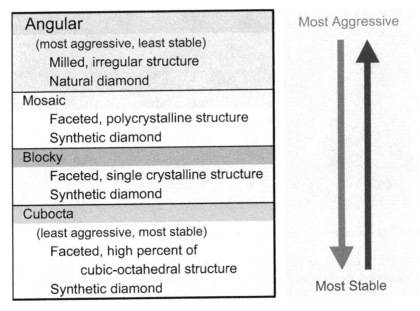

Figure 7.22 Basic diamond types with their characteristics.[13]

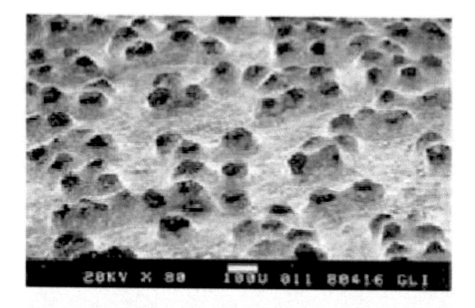

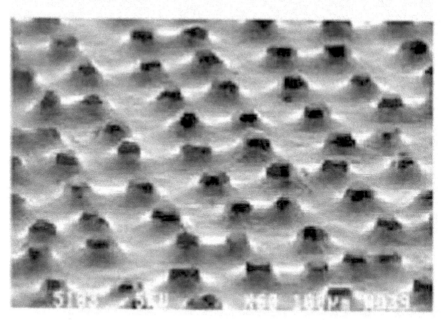

Figure 7.23 Random and structured diamond distribution patterns.[18]

might think that there is little difference in the effect of the two, as long as the diamond density is the same. This has been borne out in at least one study,[13] although it may be more probable to have a controlled distribution of diamonds with the structured manufacturing approach.

One attempt to quantify the effects of diamond profiles (including protrusion) was the development of the relative abrasive sharpness (RAS) metric.[15] The coefficient of friction is a complex combination of the surface material interactions, derived from both chemical (bonding) and physical (shape) interactions.

Pad material density could normally be used to indicate the degree of bonding or cross-linking present in a pad formulation. The presence of the pores in the material, however, makes this an unreliable index to assess material bonding. The ability of the protruding diamond to break the pad material bonds depends on the strength of the bonds and the physical (and chemical) force that is brought to bear in the conditioning process (diamond profile, conditioner downforce, relative conditioner velocity). The resistance force of the diamond–pad interaction will be a strong function of the surface condition. A surface condition dominated by the polymer pad surface is likely to be very different from one dominated by the process byproduct encrustations. It is quite possible that one diamond profile has different effects on the two surface forms. Not enough work has been done to separate the MRR performance distinction based on the extent of polymer roughness or byproduct presence. Thus, once again, application dependence plays a large role in the final process performance.

Besides the abrasive shape, the conditioning profile effect is dependent on the degree of protrusion of the diamond from the mounting surface. Figure 7.24 shows four types of conditioner disk surfaces and their relative heights.[13] However, a larger protrusion presents an increased torque on the adhesive mounting surface bond, which can lead to greater instability (depending on the adhesive method). Stability not only affects the conditioner lifetime, but can increase the probability that a diamond will come loose and become embedded in the pad. The consequences of this can be drastic, as illustrated in Figure 7.25.[19] Changes in the diamond density will lead to changes in the conditioner effectiveness, as seen in Figure 7.26.[5] The method of diamond adhesion is an important factor in diamond retention and conditioner disk lifetime.

The method of sticking the diamonds to the mounting surface forms another differentiation among vendors of conditioning disks. The typical choices are brazing, sintering, electroplating, and CVD coating. The retention capability of the various adhesion choices can be examined through retention tests. Retention capability can be assessed by measuring both diamond bond strength and diamond fracture toughness. Several methods can be employed to identify the probability of diamond retention. Two illustrated here (Figure 7.27)[14] are the high pressure water jet and the mechanical dislodging force. The water jets sweep the disk area with a high-pressure steam of water aimed at dislodging weakly bonded diamonds or presenting sufficient force to fracture weak diamonds. The surface is then scanned afterward, and the number of retention failures (and the type) per unit area is observed. The actual retention force can be assessed at sample diamond sites with a force

Figure 7.24 Conditioner disk surfaces and relative diamond height.[11]

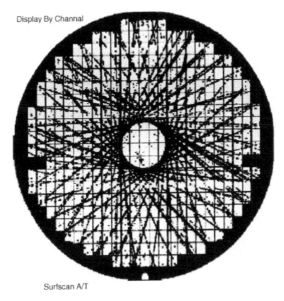

Figure 7.25 Effect of a conditioner diamond becoming embedded in the pad.[19]

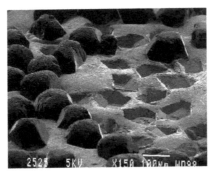

Figure 7.26 Missing diamond on the face of a conditioner disk.[5]

probe.[11] Probability plots can be made to identify trends among samples for diamond retention.

Included in the concept of the abrasive profile is the real-time change in its effective force profile due to byproduct accumulation. Various rotational and rinsing schemes have been developed to reduce the byproduct accumulation effects that would otherwise reduce the diamond's effective cutting action.

Other considerations include conditioner disk flatness (and its concomitant manufacturing control).[6] The ability to engage the diamond tips depends both on their protrusion height above the disk and the flatness of the disk surface. Manufacturing control over both elements is necessary for process stability as disks are replaced.

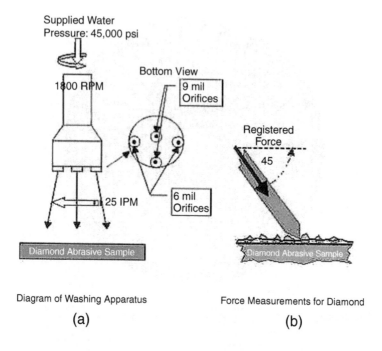

Diagram of Washing Apparatus

(a)

Force Measurements for Diamond

(b)

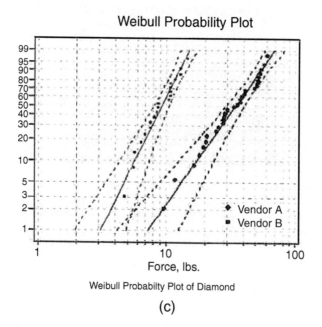

Weibull Probabilty Plot of Diamond

(c)

Figure 7.27 Force apparatus and Weibull distribution of diamond.

Ultimately, the conditioning recipe should reflect the need to return the pad to a standard material removal capability. If a high correlation can be established among directly measurable parameters (such as COF and AE) and the MRR or nonuniformity, then the process can be run only as long as the parameter readout maintains its standard or steady-state value. Many integrated circuit (IC) manufacturers run successive processes with different conditioning recipes (downforce pressure, time, sweep pattern) and check the results in restoring MRR and/or uniformity. This is often done across several different pads to even out lot variability, and then the worst-case scenario is used to set the pad conditioning recipe. The need to achieve the optimum conditioning recipe for the actual pad in use is decreased, since once the pad has been restored, further conditioning does not substantially degrade the result. Unfortunately, overconditioning does reduce pad lifetime, which will be reflected in the cost of consumables and ultimately in the profitability of the fabrication process. Studies done on the ability to set tight but realistic *unmonitored* conditioning times have shown that the conditioning recipe must often be set 2 to 2.5 times the typical pad requirements. The ability to interrupt the conditioning process when it has achieved its purpose, could reduce the pad wear rate per process step and thus extend the lifetime of the pad. The pad wear-over time is shown in Figure 7.28.

If a sensor has the ability to monitor the actual wear on the pad, then the replacement point for that pad can also be triggered by the actual conditions on the pad in use, rather than a general standard that must encompass the worst-case pad scenario.

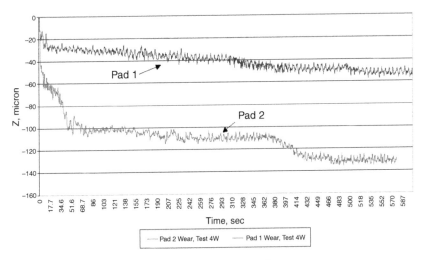

Figure 7.28 Monitoring pad wear for process control.

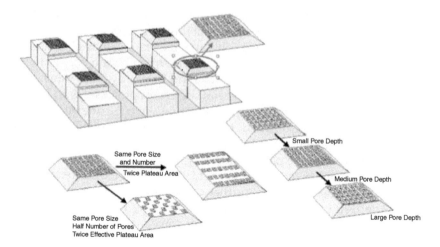

Figure 7.29 SmartPad pore combinations.[1]

The results of the FTIR analysis show no significant difference in the chemical composition between the new and the used pads, indicating no significant degradation of the used pads during polishing.[4]

The resolution of these problems may be forthcoming. What is needed is a method of separating the effects of the different contributing mechanisms. The first separation would be that of the surface abrasion of asperities versus the chemical effect of slurry availability. To do this, one would almost have to find a way to polish with the ability to change the asperity contact area and/or the slurry reservoir contact area. Their approach to solving this seemingly intractable problem is to fabricate a pad whose surface has been created *by design* through a micromachined template. The template would be able to adjust such parameters as the plateau (abrasive) area; the number, size, and density of the asperities; the abrasive area versus the pore areas; the pore volume; and so on. By carefully adjusting the ratio of the different structural parameters, much could be learned about their individual effects. An example of adjusting the pore combinations is shown in Figure 7.29.

References

1. J. Luo and D. Dornfeld, Material removal mechanism in chemical mechanical polishing: theory and modeling, *IEEE Trans. Semicond. Manuf.*, 14, 2, May 2001.
2. Y. Moon and D. Dornfeld, The investigation on the performance of chemical mechanical polishing (CMP) based upon the wafer-pad contact mode, Proceedings of the Advanced Metallization Conference (AMC), Colorado Springs, Colorado, October 6–8, 1998.
3. J. Luo and D.A. Dornfeld, Optimization of CMP from the viewpoint of consumable effects, *Journal of the Electrochemical Society*, 150, 12, G807–G815, December 2003.
4. A.S. Lawing and R. Rhoades, Pad/Slurry/Conditioning Interactions in Oxide CMP, CMPUG Presentation 12, 2001.

5. S. Qamar and T. Namola, Bond Strength and Crystal Retention Properties of CMP Pad Conditioners, Abrasive Technology, Abstract at 6th Int. Conf. CMP-MIC, Santa Clara, CA, March 7–9, 2001.

6. C. Garretson, S. Mear, J. Rudd, G. Prabhu, T. Osterheld, D. Flynn, B. Goers, V. Laraia, R. Lorentz, S. Swenson, and T. Thornton, New Pad Conditioning Disk Design Delivers Excellent Process Performance While Increasing CMP Productivity, CMP Technology for ULSI Interconnect, SEMICON West 2000.

7. R. Taylow, K. Achuthan, and A. Lucia, Complex domain distillation calculations, *Computers & Chemical Engineering*, 20, 93–111, January 1996.

8. L. Cook, Chemical processing in glass polishing, *J. Non-Crystalline Solids*, 120, 152–171, 1990.

9. J. Steigerwald, S. Murarka, and R. Gutmann, *Chemical Mechanical Planarization of Microelectronic Materials*, John Wiley, New York, 1997.

10. I. Ali and S. Roy, Pad conditioning in interlayer dielectric CMP, *Solid State Technol.*, June 1997.

11. C.-Y. Chen, C.-C. Yu, S.-H. Shen, and M. Ho, Operational aspects of chemical mechanical polishing — polish pad profile optimization, *ECS*, 147(10), 3922, 2000.

12. A.S. Lawing, Pad Conditioning and Removal Rate in Oxide Chemical Mechanical Polishing, Proceedings of the Seventh Chemical-Mechanical Planarization for ULSI Multilevel Interconnection Conference, Feb. 2002.

13. A.S. Lawing, Polish Rate, Pad Surface Morphology and Pad Conditioning in Oxide Chemical Mechanical Polishing, Proc. Fifth Int. Symp. Chem. Mechan. Polishing, ElectroChemical Society, May 2002.

14. B. Goers, G. Palmgren, and V. Laraia, Measurement and Analysis of Diamond Retention in CMP Diamond Pad Conditioners, from 3M/Semiconductor CMP/CMP Pad Conditioning (Diamond Pad Conditioning Disks), 3M Company publication, St. Paul, MN, March 2000.

15. G. Prabhu, D. Flynn, S. Kumaraswamy, S. Qamar, and T. Namola, Pad life optimization by charaterization of a fundamental pad–disk interaction property, *CMP-MIC Proceedings*, 293, 2000.

16. L. Borucki, (Motorola Inc.), Mathematical modeling of polish-rate decay in chemical-mechanical polishing, *Journal of Engineering Mathematics*, 43, 2–4, 105–114, August 2002.

17. N. Gitis, J. Jin, and D. Craven, Tribology Case Studies for Copper Removal Optimization, AVS_CMPUG101002, Center for Tribology: PadProbe.™

18. M. Bubnick, S. Qamar, T. Namola, and D. McClew, Impact of Diamond CMP conditioning Disk Characteristics on Removal Rates of Polyurethane Polishing Pads, Abrasive Technology, Abstract at 9th Int. Conf. CMP-MIC, Santa Clara, CA, February 24–26, 2004.

19. Peter Renteln, private communication, summer 2004.

chapter eight

Post-CMP cleaning

One of the major defects of CMP is the residual particles. These submicron and smaller particles cause catastrophic failure for memory disks and for ever-smaller IC chips. Any particles left behind can have a significant effect on the outcomes of downstream process steps. These include creating imperfections ranging from bumps or pits to regions of excessive electrical resistance. Post-CMP cleaning is an important process in removing these particles. This process is usually accomplished with a deionized-water or dilute cleaning agent rinse. This process involves mechanical action by contacting one or more rotating roller brushes, while the wafer or disk itself is rotated and sprayed. In order to develop effective cleaning processes to meet the stringent requirements of the fabrication of advanced integrated circuits (ICs) and information storage diskettes, it is essential to understand the mechanisms of particle removal. Figure 8.1 shows the schematic diagram of a post-CMP cleaning setup. The brush is generally made of polyvinyl alcohol (PVA). This material is open pore.

The PVA material is made into a roller shape, as shown in Figure 8.1, the Knobby brush roller. This material has a high density and is capable of absorbing up to 11 times its own weight of water. Table 8.1 shows the major properties of this material.

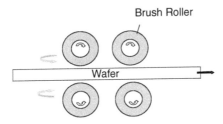

Figure 8.1 The post-CMP cleaning process.

Table 8.1 Typical Physical Properties of the Brush Used in Post-CMP Cleaning[1]

Chemical Composition	Polyvinyl Alcohol Formal (PVAf)
Porosity (%)	85–95
Average pore size (μm)	110–150
Apparent density (g/cm³)	0.70–0.110
30% compressive stress (g/cm²)	10–110
Tensile strength (kg/cm²)	2–6
Tensile elongation (%)	200–400
Water absorption rate (wt%)	700–1500
Maximum allowable temperature (°C)	80 dry, 60 wet
Decomposition point (°C)	170

Source: Rippey, *Rippey — Critical Cleaning Experts*, Technical Information, 2000.

A wafer or disk is rotating during cleaning, as shown in Figure 8.2, while the water or dilute agent is introduced between brushes. A wafer can be cleaned within one minute. The brush generally rotates at 60 rpm.

Single crystal (100) silicon wafers were used for cleaning experiments (provided by the Wacker Siltronic Corporation). The thickness of the wafers was typically 0.16 mm. The surface roughness Ra is 0.06 μm.

The mechanisms of particle removal have been studied in the past few years. Reports show that the particles adhere to a surface primarily by van der Waals forces, electrostatic attraction, or capillary action.[2] The cleaning is by hydrodynamic lubrication. The thickness of the hydrodynamic fluid layer, as estimated, was around 3.7 μm.[1] On the contrary, numerical analysis concluded that the lift force in the hydrodynamic boundary layer of fluid was too small to lift particles off the surface.[3] The possible removal force comes likely from the drag force between the brush and the wafer surface. Major

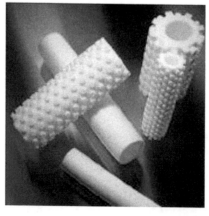

Figure 8.2 Knobby brush rollers. (From the Rippey Corporation.)

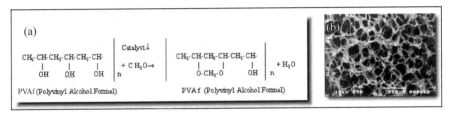

Figure 8.3 (a) Reaction formula of the PVAf and (b) pores of a PVAf foam.

research areas in cleaning include comparing the influence of the design of the brush on the efficiency of cleaning, the influence of chemistry on the cleaning effectiveness,[4–9] and the influence of the chemical properties of the solution used to soak the brush on the physical properties of the brush.[1]

As discussed in Section 4.4, the post-CMP cleaning is a tribological process that involves three bodies rubbing against each other. Figure 4.18 shows the estimated friction coefficient against the sliding speed using a laboratory system. This figure has two curves: one is for a dry or moist brush and the other is for a brush soaked with water. The wet brush apparently has a lubricating behavior as the friction decreases with speed. Because of a lack of lubrication, the dry brush on the contrary has an increasing friction against speed. In order to study further the lubricating behavior of the wet brush, we decided to investigate the wet brush using the renowned Stribeck[10] approach, which has been used for nearly 100 years. Figure 4.19 shows the friction coefficient on the ordinate and the Sommerfeld grouping (viscosity × speed/load) on the abscissa. In this figure, the friction coefficient decreases as the load increases. When the Sommerfeld grouping increases, the friction coefficient lowers more quickly for higher loads (1713 g and 3549 g) than for lower ones (394 g and 472 g). For low loads, the friction coefficient remains almost stable after a certain grouping number.

The Stribeck curve[11] shows the evolution of the friction coefficient (μ) versus the Sommerfeld grouping ($\eta \cdot U / W$), where η is the dynamic viscosity, U is the surface speed, and W is the load. This allows us to define the lubrication regime involved in a lubricating system.

According to the standard Stribeck curve shown in Section 4.4 (Figures 4.18 and 4.19), we can identify the lubrication regime for a cleaning system: PVA and silicon, and in the range of speed and load considered. The shape of the curve is first decreased in its friction coefficient, which corresponds to boundary lubrication. Then the friction coefficient remains constant, which corresponds to the mixed lubrication regime. There is no hydrodynamic lubrication regime because the friction coefficient is much higher than the known value for a hydrodynamic friction for water, i.e., 0.02.[12]

The contact is illustrated in Figure 8.4. The figure indicates the multiasperity contact throughout the cleaning process. The modeling continues to examine the behavior of one nodule on the brush. Five points on one nodule were modeled, as shown in Figure 8.5.[13] During contact, the sliding occurs

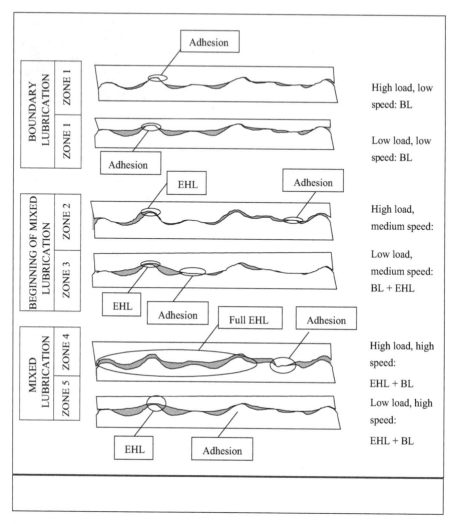

Figure 8.4 Description of the different lubrication regimes.

predominantely. Interestingly, however, adhesion happens occasionally. The contact shear stresses and the friction forces at five points on the contact surface change with time. This is shown in Figure 8.6. In Figure 8.6a, the top point is first in contact with the wafer and the bottom point is the last. The five points shown in Figure 8.6b share a similar feature, i.e., contact shear stress increases to a peak and then drops to zero when contact is lost. The top point, however, has the highest value, about 16 kPa, as it is the leading edge of the nodule. The initial contact shear stress of the bottom point has the opposite sign from that of the rest of the points; this is caused by adhesive (not sliding) friction. When sliding occurs, the normal stress can be found by dividing the shear stress by the coefficient of friction. For a coefficient of friction of 0.2, the normal stress is five times the shear stress.

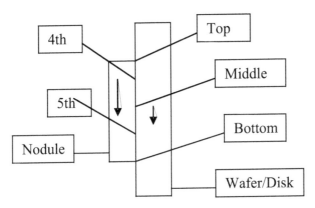

Figure 8.5 Brush nodule for modeling.

The total shear force on the contact boundary between the nodule and the wafer is shown in Figure 8.6b. The shear force is found by integrating the shear stresses over the entire contact area. The shear force rises from zero to a peak value of about 0.23 N and then drops to zero. Here we have seen that even though experimentally measured friction is stable at a constant value, the real contact forces between each nodule and wafer or disk surface actually change owing to different locations and surface geometry.

The strength of a chemical bond is often known as *the bond dissociation energy*. It is defined as the standard enthalpy change of the reaction in which the bond is broken. Let us assume that a silicon wafer is to be cleaned. The slurries used for CMP have either Al_2O_3 or SiO_2. There are most likely four types of chemical bonds involved in chemical bonding, as shown in Table 8.2.[14] In this table, we also list the value of bond length. Here we do not consider the angle or the orientation of the bonds.

From the data given, the forces needed to break these bonds are therefore in the range of 1.81×10^{13} N to 2.99×10^{13} N. If we assume that the particles adhere via van der Waals forces, however, typically the strengths for intermolecular forces lie in the range of 0.24 to 2.39 cal/mol. The forces needed to break the bonds between particles and a silicon wafer are more than ten orders of magnitude higher than with the measured and calculated friction force shown above, so it is clear that the energy provided through friction is not enough for the removal of particles. This leads to several possibilities:

Owing to the lack of direct mechanical force acting on the particles, they are not removed through one-pass abrading or the scratching effect.
Mechanical stimulation triggers the chemical dissolution of certain single-crystal metals, as has been reported by Dickinson.[15] Therefore, in cleaning, the continued mechanical impact through brush or liquid on particles will activate the bonds between particles and the wafer surface.

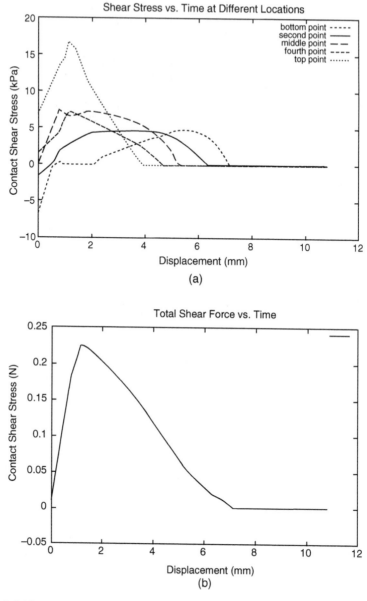

Figure 8.6 Numerical analysis showing stress distribution on a brush nodule.

Chemical interaction is again important in the cleaning process. Weakening the bonds yields effective particle removal.

The understanding of the post-CMP process is still in the beginning stages. We hope that concepts discussed in this book will help engineers to know how to solve certain problems.

Table 8.2 Bond Strengths of Potential Particles with Si
Surface[14]

Bond Type	Strength (Cal/mol)	Bond Length (Å)	Remarks
Si–Si	78.1	2.3	
Si–O	191.1	1.5	Silica slurry
Si–Al	54.8	—	Alumina slurry
Si–H	71.5	1.4	

References

1. Rippey Corporation, Rippey — Critical Cleaning Experts,™ Technical Information, 2000.
2. D.W. Cooper, R.C. Linke, and M.T. Andreas, Comparing the effectiveness of Knobby and ridged post-CMP cleaning brushes, *MICRO*, July/August 1999.
3. D.S. Rimai, D.J. Quesnel, and A.A. Busnaina, The adhesion of dry particles in the nanometer to mircrometer-size range, Colloids and Surfaces A: Physicochemical and Engineering Aspects, 165, 3–10, May 30, 2000.
4. E. Zhao, L. Zhang, H. Li, D. Hymes, J.M. de Larios, and W.C. Krusell, Copper CMP Cleaning Using Brush Scrubbing, Technical Proceedings CMP–MIC Conference, San Jose, CA, 1998.
5. G.S. Higashi and Y.J. Chabal, in: *Handbook of Semiconductor Wafer Cleaning Technology*, Werner Kern, Ed., Noyes Publications, Norwich, NJ, 1993, pp. 433–496.
6. D.J. Hymes, M. Ravkin, X. Zhang, and W.C. Krusell, Method and apparatus for cleaning of semiconductor substrates using hydrofluoric acid (HF), U.S. Patent 5,868,863, 1999.
7. J.M. Steigerwald, S.P. Murarka, and R.J. Gutmann, *Chemical Mechanical Planarization of Microelectronic Materials*, John Wiley, New York, 1997.
8. D.W. Cooper, R.C. Linke, and M.T. Andreas, Comparing the effectiveness of knobby and ridged post-CMP cleaning brushes, *MICRO*, July/August 1999.
9. M.A. Ravkin, D.L. Hetherington, J.M. de Larios, D.G. Gardner, and W.C. Krusell, A New Chemical Mechanical Scrubbing Process Using HF for Post-CMP Cleaning Application, Technical Proceedings CMP–MIC Conference, San Jose, CA, 1996.
10. R. Stribeck, Characteristics of plain and roller bearings, *Zeit. Ver. Dent. Ing.*, 46, 1341–1348, 1902.
11. C.A. Coulomb, *Mémoire de Mathématique et de Physique de l'Académie Royale*, Paris, 1785.
12. Prof. A. Mori and Prof. H. Liang, private conversation, July 2000.
13. H. Liang, E. Estragnat, J. Lee, K. Bahten, and D. McMullen, Mechanisms of Post-CMP Cleaning, Proc. Sixth Int. Conf. CMP for ULSI Multilevel Interconnection (CMP–MIC), Institute of Microelectronics Inter-Connection, March 7–9, 2001, Santa Clara, CA, pp. 266–272.

14. *CRC Handbook of Chemistry and Physics*, 68th ed., CRC Press, Boca Raton, FL, 1987–1988.
15. J.T. Dickinson, N.-S. Park, M.-W. Kim, and S.C. Langford, A scanning force microscope study of a tribochemical system: stress-enhanced dissolution, *Trib. Lett.*, 3, 69–80, 1997.

Index

O

Oil viscosity, 63
    effects of, 78
Open-cell elastomeric foams, 121
Optical profilometer, 48
Optical surface characterization techniques,
    48
Optimized conditioning patterns, 152
Orbital CMP tools, 27, 132
Outlet region, 74
Overconditioning, 165
Overpolishing, 49
Oxide film, 45
Oxides, adhesion of, 83

P

Pad slurry distribution support geometry,
    147
Pads, 25, 30, 63
    asperity-rich surfaces, 135
    conditioning, 26, 64, 131, 144
    fatigue wear on surfaces of, 93, 96
    glazing, 152
    interactions with wafer, 79
    life of, 96
    material density, 161
    polymeric, creep and relaxation in, 110
    role of, 101
    surface characteristics, 135
    surface profile, 137
    wear during polishing, 98, 154, 165
Particle removal, 169
Passivation, 45
    copper, 81
    metals, 79
    surface, tribochemical wear and, 96
Pattern density effect, 32
Peak-to-valley height, 39
Penetration hardness, 43
Petroff's law, 78
Physical adsorption, 45
Physical defects, 49
Physisorption, 45
Pits, 49, 169
Pitting, 96
Planar processing, 8
Planarization
    copper, temperature fluctuations in, 27
    definition of, 1, 33
    efficiency, 32
    length, 32

Plastic buckling, 123
Plastic collapse deformation mode, 121
Plastic collapse stress, 125
Plastic deformation, 42, 95, 137
    closed-cell structures, 127
    Hertz's equation for, 74
Plateau length distribution, 118
Ploughing, 95
    friction due to, 59
Poisson's ratio, 73, 103
Polar lubricant molecules, 45
Polishing
    conditions and wear, 98
    copper, 81
    definition of, 1
    force transmission and, 101
    glass, 83
    performance, 71
    removal of byproducts of, 140
    tungsten oxidation rates during, 80
    urethane pad, 64
Polymer pad, 25, 30
Polymer pores, 115
Polymers
    friction of, 64
    mechanical properties of, 101
    viscoelastic behavior of, 105
    wear in, 98
Polyurethanes, 121
    deformation of, 128
Polyvinyl alcohol (PVA), 169
Pore effect, 132
Pore structure, 101, 135
Pores, 115
    distribution and density of, 117
    trailing edges, 131
Post-CMP cleaning, 169
    lubrication behavior in, 87
Potassium dispersal of nanoabrasive
    particles, 81
Potentiostatic slurry control, 80
Pourbaix diagram, 72
Precision interferometric optical metrology
    systems, 139
Pressure-viscosity coefficient, 45, 75
Preston's equation, 23, 145
Principal shearing stress, 44
Probability springs, 101
Process conditions, 84
Profilometers, 48
Profilometry, 40
Proportionality constant, 58
Protruding diamonds, 161

Milton Keynes UK
Ingram Content Group UK Ltd.
UKHW040056071024
449327UK00019B/605

9 780367 393250